Development, Environment and Sustainable Livelihood

Development, Environment and Sustainable Livelihood

Edited by

Soumyendra Kishore Datta
and
Atanu Sengupta

CAMBRIDGE SCHOLARS

PUBLISHING

Development, Environment and Sustainable Livelihood,
Edited by Soumyendra Kishore Datta and Atanu Sengupta

This book first published 2014

Cambridge Scholars Publishing

12 Back Chapman Street, Newcastle upon Tyne, NE6 2XX, UK

British Library Cataloguing in Publication Data
A catalogue record for this book is available from the British Library

ISBN (10): 1-4438-6305-X, ISBN (13): 978-1-4438-6305-6

TABLE OF CONTENTS

Section D: Natural Resources and Livelihood

ACKNOWLEDGEMENTS

This book is an outcome of an international conference held in the Department of Economics, Burdwan University in 2013. The broad area of conference had been related to development, environment and livelihood issues which are also in some way linked to ongoing DRS project theme in the Department, pertaining to issues on rural livelihood. The book is based on selected papers presented in the conference and covers various subthemes like Farm sector and livelihood issues, Gender issues, Socio-developmental issues, and Natural resource and livelihood. Since, now-a-days the efforts related to development are considered as some sort of threat to the maintenance of environmental quality which may have far reaching repercussions on livelihood aspects, major focus in the conference has been put on these linked issues. We are extremely grateful to the UGC and Burdwan University authority for providing the financial support to hold the conference. We really extend our thanks to all the authors who have contributed to the shaping of this book. During the course of editing of this book we were also benefitted by discussions and interactions with Prof. Arup Chattopadhyay, Head, Department of Economics, The University of Burdwan. We are also thankful to other faculty members including Pinaki Chakraborti, Maniklal Adhikary, Rajarshi Majumdar, Bhaskar Goswami, Jhilam Ray, Sita Lama who did not hesitate to extend their cooperation, whenever required, during the process of this work. We also extend our sincere thanks to our Vice Chancellor, Prof. Smriti Kumar Sarkar who always motivated us to undertake this sort of academic venture. It would be a serious omission if we don't recognise the untiring efforts of research students Tanushree De and Krishna Singh who never wavered to offer help and support during the entire process of editing this book and we thank them as well. Last, we feel obliged to acknowledge the efforts of editorial and publishing staff at Cambridge Scholars Publishers, U.K. Despite this the responsibility of any errors of omission and commission that may remain solely rests on us.

Soumyendra Kishore Datta
and
Atanu Sengupta

INTRODUCTION

The achievement of higher economic growth is one of the principal objectives of the government policies. It is a great challenge for the developing countries and only vehicle which can bring these countries out of poverty. This also involves resource intensive development programmes with equitable access and distribution of output. Sustainable human development is conditioned by the equitable distribution of and access to such output and enhancement of human capabilities. Aspects of human development, have for a considerable period of time, been the central issue in the development debate for the world economies, especially the third world countries. The concern for making provision for better opportunities and advantages for a reasonably good standard of living for the majority of human beings, if not all, has been expressed in different quarters in different forms.

Apparently it seems that the countries with higher average income tend to enjoy higher average life expectancies, lower rates of infant and child mortality and higher literacy rates and hence a higher human development index. But inter-country comparisons in this field reveal that these income variations tend to explain less than half the variations in life expectancy, infant/child mortality and differences in adult literacy. The central focus of this statement is that in assessing human development the matter of concern should not be income alone but the use that this income is put to. A society can spend its income on arms without expending on education. An individual may buy narcotic drugs without consuming essential food.

Thus economic growth is very important but at the same time it is insufficient for human development. Hence enhancement of human capabilities rather than promotion of aggregate growth of gross national product alone should be the main concern.

Sometime back it was rife among well informed people that along with economic growth, through improved technology there will be availability of proper substitutes and there will be an efficient use of the existing natural resources.. But mere efficiency in the use of natural resources in the present generation does not guarantee its availability in future generations as well. There are certain ecosystem functions like air and soil quality, climate regulation, disturbance regulation, nutrient cycling, waste treatment etc which are quintessential for the existence of human race and cannot be replaced. Thus in order to make economic development sustainable, natural and environmental resources must be used in such a way that there is enough for the present generation as well as the future generation.

Human development is also intimately related to access to the aforesaid type of resources and how these resources are distributed and utilized in shaping peoples' livelihood. India's development path is based on its unique resource endowments. As a welfare state the overriding priority lies in generating peoples' wellbeing with the multifarious programmes of eradicating poverty through providing means of earning income for a sustainable livelihood.

There is no counter to the fact that despite elapse of 60 years since independence the teeming millions in the Indian economy still reside in country side where fight against poverty for many still remain a regular phenomenon. Govt. has adopted a number of poverty alleviation strategies from time to time to provide sustainable earning opportunities to these people and raise their income and consumption level. While these are generally targeted to achieving the goals of eliminating rural poverty, the purpose of generating an all-round enhanced livelihood opportunity based on creation of an improved ambience is only partially served by such programmes. The recent focus, therefore, has been on the assets/processes/ activity framework concerned with not only poverty reduction but also promoting sustainable livelihood enhancing strategies and access to assets like human capital, physical assets, social capital, financial capital and natural capital. In terms of the sustainable livelihood framework a livelihood comprises the activities, the assets, the capabilities and access that combine together for determining the living attainable by an individual. A livelihood is deemed to be sustainable when it can absorb unforeseen shocks and recover from the stresses and uncertainties while maintaining or enhancing the capability and asset base both at present and for future periods without distorting the natural resources and creating social unrest. There are seventeen papers in this book covering socio-developmental aspects as well livelihood issues intimately linked with the Farm and Non-farm Sector, natural resources as well as influenced by gender issues.

The paper by Arpita Banerjee and Pravat Kumar Kuri makes an attempt to examine the growth performances of agricultural production and productivity of major States of India and the level of disparity in the performances of agriculture since 1970-71. The paper also explores the nature of cropping pattern in India and its states over the study period. The growth performances have been analysed considering three distinct phases of agricultural development in India viz the first phase of green revolution 1970-71 to 1979-80, second phase of green revolution 1980-81 to 1990-91, and the period after economic reform 1991-92 to 2007-08). The variability in agricultural output is observed to be responsible partly due to the variability in agro-climatic conditions across the states and partly due to the variations in agricultural infrastructure. Moreover, using cross section and time series analysis, this paper attempts to examine the trends of convergence/divergence of per capita value of agricultural output over the period 1970-71 to 2007-08. The results of tests of σ and absolute β convergence show a diverging tendency in agricultural growth in India. The unit root test result identifies the states that are responsible for divergence in agricultural growth in India.

The paper by Biswajit Ghosh and Namita Chakma used the notion of ecological footprint and biological capacity as indicators for measuring regional sustainable level of land use in West Bengal. They used the applied component method which was used by (Li et al, 2009) in assessing ecological footprint of land use in China, 2009. The content of the work is divided into two parts: the ecological supply (or bio-productive areas) and the demand on nature (or ecological footprint). They found that some of the districts like Darjiling, Malda, Purulia, Bankura, Haora and South 24 Parganas were endowed with less cultivable land and low cropland bio-capacity (less than 1000000 ha), while districts like Dakshin Dinajpur, Murshidabad, Nadia and Paschim Medinipur had large amount of cultivable land high cropland bio-capacity. Ecological surplus were enjoyed by districts having more bio-capacity of cropland than footprint while the reverse happened to be the case with ecological deficit, districts suffering from more footprint of cropland than bio-capacity. They concluded the paper with the findings that 14 districts were experiencing ecological deficit while nine districts amongst them were having a very crucial condition. There include Malda, Murshidabad, Nadia, North 24 Parganas, Haora, Hugli, Barddhaman, PurbaMedinipur and Kolkata which were afflicted with a high density of population.

Nirmalendu Sarkar, Santosh Kumar Dutta and Swapan Kumar Biswas, in their paper ventured to assess the remunerative crop or combination of crops in an agricultural year of any cropping system suitable to a specific agricultural region, by considering both production costs and annual net returns. . The study undertaken in the context of some selected villages in the district of Budwan, West Bengal had been based on the fact that improving the productivity of agricultural land largely depends on the comparative advantage of agriculture of any cropping system selected by the respondent farmers. It is observed by them that operation and interaction of a variety of factors like production cost in terms of seed and fertilizer requirements, expenses on modern implements, size of land holding and usability of machine, electricity used for agricultural purpose, marketing infrastructure and transport facilities, are usually crucial in influencing the productivity and profitability in agricultural scenario.

Arup Majumdar in this paper, has tried to present the results of an anthropological field based study among a group of peasant families in the villages under Kharagpur- I Block in Paschim Medinipur district of West Bengal. He specifically put focus on the comparative conditions of the women of landloser and non- landloser families subsequent to the acquisition of agricultural land for the establishment of a heavy industry. He showed that after the acquisition, livelihood pattern have changed among the female members of landloser families.

The paper, "Microfinance Access And Women Empowerment In India: An Inter-State Analysis" written jointly by Arindam Laha and Pravat Kumar Kuri deals with the problem of microfinance outreach in enhancing the economic opportunities among the women. The paper delineates two aspects: the microfinance outreach and women empowerment. Both these aspects are collated into two separate indices that are a combination of several dimensional indicators. The authors then look at the relation between these two indices. Interestingly a positive association is established between microfinance outreach and women empowerment. The states having higher level of microfinance outreach are also the states with a relatively high level of women empowerment. It is, thus, predicted that an all-inclusive microfinance system would strengthen the process of financial inclusion in India and thereby would promote women's empowerment.

Soumyendra Kishore Datta and Tanushree De focus on the relative discrimination and deprivation of females compared to male counterparts in the sphere of diversification in job market and find that the differential in livelihood diversification is well correlated gender differential in capability. They also ascribe this female discrimination to observed differences in asset ownership and mobility status across gender. They report that in the developing regions, they still have far less access to material resources, education and training, health related benefits and economic and employment opportunities. Particularly they remain deprived in terms of provision of health and education services for women. There operate several constraining factors that continue to keep their capabilities at a subdued level compared to males and in the process females usually do not get opportunities to pursue a broader diversified form of livelihood unlike their male counterparts. This often stands in the way of their continuous source of earning and affects financial independence and social dignity. Sometimes the gender differential in socio-economic attainments is so acute that women face severe problems in earning from a diversified access to work opportunities.

Soumitra Sarkar has considered the problem of women empowerment when they seek support through multiple borrowing from micro finance institutions. He undertook a case study in the Rajganj block of Jalpaiguri district. It focuses on the fact that often when it comes to reporting during project evaluation stages, figures show that repayment is meeting recovery targets. But beyond the recovery factor there exist some chains of financial transaction which govern the efficacy of functioning of SGSY scheme and have bearing on women empowerment .Citing some case study examples he shows how some women members of certain SHGs fell into debt trap while trying to repay the interest and principal of one loan by taking recourse to a second and sometimes even a third loan from various different private .financial institutions. Data collected from field survey revealed that beneficiaries often undertook debt swapping which had indeed been a serious phenomenon that proliferated subsequent to emergence of microfinance systems in the area under the study.Excess debt intake often results due to absence of adequate knowledge of the debt schemes i.e., term of debt, conditions of debt, rates of interest etc. According to him unless the phenomenon multiple lending/ multiple borrowing be stopped the beneficiaries would not be able to reap the full benefit of the microfinance system and the aspect of women empowerment would be continue to be eluded.

Utpal Kumar De has tried to focus attention on relative efficacy of the look east policy undertaken by the Govt. of India in the event of globalization for more regional and sub-regional cooperation and understanding for the promotion of trade, especially export in the eastern and south-east Asian countries. Attempts have been made in this paper to show how far India has been able to integrate with other Asian countries in terms of various globalisation indices and expand foreign trade with these nations and finally took advantage of open policies to accelerate the growth of GDP and HDI. The study came out with the conclusion that despite the emergence of a notable crisis in the fast growing Asian countries after 1997 and their subsequent recovery, India's effort to catch up the eastern markets has not been highly successful. In terms of growth in exports India could not fare well commensurate with those of other countries and the trade deficit with major ASEAN countries still continues.

Atanu Sengupta and Debjyoty Mukherjee in their paper investigated whether performance of individual physicians within the same "Service-State" differs and if yes how? They define service-state as a compound of different dimensions that includes stock of infrastructure, consumables, quality of support staff and number and attitude of patients.They record self-reported performance of physicians within a given service-state considering knowledge asymmetry of licensed physicians has a downward rigidity. Mainstream economists developed a measure of objective efficiency-the standard stochastic, DEA and other such methods. In trying to look at mathematical precision, economists lose much simple truth that lie underneath such complex analysis. This paper goes beyond the gamut of traditional efficiency measures and brings the conception of measures through subjective efficiency.

Debasish Batabyal analysed the prospect of Tourism development in Sikkim which has new dimensions in many respects. According to him, the growth and development potentials of tourism in Sikkim, is usually measured only with arrival statistics over the years with no available information about the origin of tourists inside and outside the country. He deems it important to reconsider the development potential of Sikkim with the emergence of problems like over-crowded routes and resulting negative socio-cultural and environmental consequences. This can be possible with sector specific focus on the tourism industry e.g. accommodations, attraction features, transportation etc. The present study is a sector specific analysis of performances of tourism elements in Sikkim with the help of WES tourism index with relevant weights. This WES tourism index has covered two aspects of tourism industry viz. data collection for each and every sub-sector in tourism and measurement of performance (Bruges and hinterland area, West-Flanders region in Belgium; 1962). He found that there had been variations in tourism index over the years under consideration. Finally he concludes that the types of tourism, types of tourists, spending pattern, alternative routes and channelization of tourist traffic are found to be the core areas of development and management in Sikkim tourism with its regional disparity in tourist arrival and varied development of supply components.

Anish Kumar Mukhopadhyay stresses that since its introduction, the choice of variables and the methodology of construction of Human Development Index, has given rise to great controversy. The UNDP made some drastic changes in the Human Development Report- 2010 by addressing certain issues pertaining to the index in order to engineer quite substantial redressal of the previous shortcomings. He mentions that changes have taken place in the field of gender bias with the advent of new indices in place of the earlier ones. However these new indices are also not without criticism. Arguments are pertinent and calls for improvements especially in the question of inequality both for Human Development and Gender. This paper investigates the actual scenario with some greater details both in terms of theoretical construction as well as from empirical point of view in a cross-national framework.

Soumya Sahin and Ambar Nath Ghosh constructed a simple overlapping generations model in which parents are assumed to derive utility from their children's education. They also used this framework to understand and analyze the implications of trade policies on skilled vis-à-vis unskilled wage and also on the relative sectoral sizes. The production sector is divided into two parts: an export sector and an import competing sector. The former produces only for the export market, while the latter caters only to the domestic market and its output competes with imports.Labour is considered as the only factor of production. While only adult unskilled labour and child labour are assumed to be employed by the export sector, the import competing sector uses both skilled and unskilled labour. Further on assumption of competitiveness the export industry is viewed as earning only zero-profit. The country under consideration is assumed to be a small open economy and therefore faces a fairly elastic demand. Then, a rise in the tariff rate will raise skilled wage, and thereby increase inequality. Also the increase in tariff rate leads to expansion of the import competing sector and contraction of the export sector. The incidence of child labour depends solely on the international price of the home country exportable.

Tonmoy Chatterjee and Kausik Gupta in their paper attempt to develop an integrated framework covering aspects of international fragmentation, trade liberalization and health sector. In order to attain this theyconsidered two different models based on Heckscher-Ohlin-Samuelson general equilibrium structure, with special stress on the health sector. The first model is devoted in considering four sectors and it was assumed that the production process of the health sector could be fragmented. In this structure an analysis was made about the impact of movement from no fragmentation regime to a regime of fragmentation on the output levels of health sector. They observed an expansionary effect on the health sector through shifting to a fragmented regime.In the second model they considered three sectors and assumed the production process of the health sector as fragmented. In this framework they elicited the result that trade liberalization would lead to an enhanced output level of the health sector.

Ruma Kundu in her paper seeks to analyse the prospect of organic farming in providing sustainable livelihood in Sikkim and its potential in empowering rural women. She undertook field survey in two different parts of East Sikkim. While the growers in Samdur operate on an individual basis, those in Sajung consist primarily of woman cultivators working and sharing profits on a group based system. The group cultivation system has demonstrated a remarkable degree of success in this newer method of cultivation. This paper uses the regression methodology for analysing both cases. She has also thrown some light on the customers' willingness to pay a premium for organic products and the possible determinant factors by using logistic regression. Apart from this, she has also examined the possible environmental benefits from the use of organic products by employing the Contingent Valuation Method.

The paper by Pradip Kumar Parida and Notan Bhusan Kar deals with an important dilemma in the development programs in India. The arena of their analysis is Chhattisgarh formed as a separate state to safeguard the interests of the inhabitants in that land . However vigorous development policies carried out in the state have produced large scale evacuation and relocation. The protest of these development refugees are then treated as anti-developmental and detrimental to the interests of the state. The state is coming forward to create another movement to stop the genuine movement and declare it as a conflict zone. What about the 'life' and 'livelihood' of a citizen from the point of view of 'human security' in a conflict zone. It tries to examine the approach to development strategy as well as the voice of people in that context. It also tries to find out where does the 'subalterns' stands today in the Indian state whose constitution categorically mentions about establishing the so called 'Socialistic Pattern of Society'. What about ecology of the society and the involvement of the community in terms of maintaining their livelihood vs. the economic development of the state and the security of the individuals, i.e. - life & liberty.

Banani Ghosh and Swapna Ghorai in their paper focus on the importance of sustainable forest management practices which has the capacity to influence the socio-economic conditions related to livelihood of the forest fringe communities. According to them, sustainable livelihood through forest management ensures more income, increased well being, reduced vulnerability, improved food security and more sustainable use of natural forest resource base that continues to provide similar benefits and productivity in the future as well. They have tried to analyse the intensity of forest based sustainable livelihood in the context of forest fringe communities of Jhargram Block situated in the western part of Paschim Medinipur district of West Bengal. An attempt has been made to develop strategies for both sustainable livelihood using forest resources and also the conservation & management of forest resource. The aspect of forest degradation and land use map has also been provided with the help of RS & GIS.

Suchismita Mondal Sarkar focused on the issue of livelihood diversification in the context of selected villages in the Burdwan District of West Bengal. Only primary data were collected and analysed for arriving at results and conclusions. She found varying degree of diversification across the villages depending on socio-economic differences of rural people as well as their attitude towards diversification. However she also noted several constraints faced by the rural households in the study area. Most of them are socio-economic in nature and can be overcome if proper govt. steps are taken. Though Government of India is running many wage employment programmes, the performance these schemes are not encouraging. Many of these programmes have not much relevance to local resources, needs and priorities. She pinpointed the most important hurdles in the implementation of such programmes, embedded in illiteracy and indifference of the rural people towards the development schemes , lack of resources, wrong selection of beneficiaries, local dominant culture, factionalism, lack of identification of felt needs of local people, etc

SECTION A:

FARM SECTORS AND LIVELIHOOD ISSUES

CHAPTER ONE

AGRICULTURAL GROWTH AND CHANGES IN CROPPING PATTERNS IN INDIA: AN INTERSTATE ANALYSIS

ARPITA BANERJEE AND PRAVAT KUMAR KURI

1. Introduction

The Agriculture sector in India has been the predominant sector as it provides the livelihood of nearly 60% of the workforce in India. In spite of the slowing down of its contribution, agriculture still contributes around 14.2% of GDP in the country, and its role is essential in promoting inclusive growth, enhancing rural income and sustaining food security. Undoubtedly, agriculture has growth linkages with other sectors of the economy. Empirical studies show that a unit increase in agricultural output would have a positive effect on both industrial production and national income and thus, have significant implications to widespread inequality in per capita income in India. In India there exists a wide regional variation in terms of agro-climatic conditions, persistence of rainfall, resource base, irrigation facility and infrastructural development, and accordingly there is wide regional variation in the performances of agriculture. In fact, there is a plethora of literature about the growth, instability, cropping patterns and interstate differences in agriculture in India. Several studies such as Rao (1975), Mehra (1981), and Desai & Hazell (1982) have pointed out that the new strategy of agricultural production based on HYV seed fertilizer technology has contributed to the growth in production and productivity in India. In an in-depth study, Bhalla & Singh (2009) showed that there had been a marked acceleration in both the output and yield growth rate in agriculture during 1980–83 to 1992–95. This result was supported by many authors like Sawant & Achuthan (1995), who claimed that the yield rate of both food grain and non-foodgrain crops accelerated significantly along with output growth during the 1980s. However, these studies also observed marked deceleration in the level and trend growth rate of output and yield during the 1990s. Several attempts have been made in the literature to identify the factors causing deceleration trends in agricultural growth in India. Regarding the convergence/ divergence of agricultural growth across Indian states, the study by Kalirajan et al. (1998) found a long-term divergence and cyclical pattern in agricultural growth. Ghosh (2006) also found absolute divergence and conditional convergence in agricultural growth in India. Under this backdrop, this study analyses the trend and disparity in the rate of growth of value of output and yield in agriculture in India since the 1970s with explicit focus on the phases of green revolution and new economic reform. The study also focuses on the nature of cropping patterns in India and its major states. Finally, a convergence/divergence test has been done with respect to the per capita value of agricultural output to find the nature of convergence/divergence in India and to identify the states responsible for causing convergence/ divergence in agricultural growth in India. For convenience, the chapter is divided into four sections. Section 2 deals with data sources and methodology used in this study. Section 3 discusses the results obtained in this study in respect of growth rates, cropping patterns, construction of agricultural indices and the convergence tests. Section 4 presents the concluding remarks.

2. Data Sources and Methodology

2.1 Data sources

The period of the study covers thirty-eight years from 1970–71 to 2007–08. Twenty major states have been studied for interstate comparisons. The study uses secondary data exclusively. The state wise and crop wise values of output for the period 1970–71 to 2005–06 have been taken from different issues of CSO, a Govt. of India publication. The figures for the value of the agricultural output of Indian states have been converted at a constant price, using 1999–2000 as the base year. Eight major crops have been selected for this study: rice, wheat, jute, cotton, sugarcane, rapeseed, mustard and potato. The state wise data on inputs and operated areas for different years are collected from different issues of Indian Agriculture in Brief, Directorate of Economics and statistics, Ministry of Agriculture and Centre for Monitoring Indian Economy, Agriculture etc. The data for area, yield and production of some selected crop for India and the states are collected from different issues of the Centre for Monitoring Indian Economy and Agriculture.

2.2 Methodology

The statistical and econometric methodologies followed in this study can be summarized under three headings relating to the estimation of growth, construction of agriculture infrastructure index and convergence analysis. For the computation of growth rates of the value of agricultural output and productivity, the trend or exponential and kinked growth rates have been computed. Changes in cropping patterns overtime have been measured in terms of relative change in area under crops to the changes in GCA. For measuring the disparity among the states in respect of different agricultural infrastructural indicators, the Composite Index of Agricultural Infrastructure is computed by using principal component analysis. The composite indices are computed using the deprivation method. To construct this index, eight agricultural development indicators are selected: Cropping intensity (CI), Percentage irrigated area to GCA (IAGC), Fertilizer consumption per hectare of GCA (fcgc), Credit to agriculture (CTA), Number of tractors and pumpsets used per 1,000 hectares (TAP), Average yield of agricultural land (AY), Road length per 100 sq km (RL), and Percentage share to total consumption of electricity in agriculture (CELA).The steps taken for computation of the index are: in the first place the component indices are constructed by using the formula $I_{ij}=(X_{ij}-\min X_{ij})/(\max X_{ij}-\min X_{ij})$, where I_{ij}=component index for the jth state with respect to the ith variable, X_{ij}=actual value of the jth state in the ith variable, and Min X_{ij} and MaxX_{ij} are the minimum and maximum values of the ith variable. Subsequently, using the Principal Component Method, weights are determined for different indicators used, and finally the respective index is computed by using the formula

$$I_j = \frac{\sum_{i=1}^{n} W_i I_{ij}}{\sum_{i=1}^{n} W_i},$$

where wi=weight attached to ith indicator, and I_j=the index of the jth state. For convergence analysis, σ convergence and absolute β convergence have been tested in the cross sectional framework and a time series analysis(unit root test) has been made to identify which states are responsible for the convergence/ divergence of agricultural growth in India.

3. Results and Discussion

3.1 Growth performance of agriculture in Indian states

The growth performance of the states has been analyzed considering three distinct phases of agricultural development in India: the first phase of the green revolution (1970–71 to 1979–80), the second phase of the green revolution (1980–81 to 1990–91), and the period after economic reforms (1991–92 to 2007–08). The exponential growth[1] rates of value of output and productivity for the country during the whole period (1970–71 to 2005–06) reveal that all Indian agricultural output and yield grew at the rate of 2.6% and 1.43% respectively. The states like Haryana (3.46%), Madhya Pradesh (3.38), Rajasthan (3.36%), Goa (3.3%) and West Bengal (3.24%) achieved higher growths in the level of output, whereas the states like Madhya Pradesh (2.81%), Rajasthan (2.5%), Haryana (2.5%), Punjab (2.12%) and Uttar Pradesh (2.04%) registered higher growth in productivity during the entire period. The performance of the state Jammu and Kashmir remained poor in both respects during the whole period. The study revealed that the performance of the nation during the first phase of the green revolution (1970–71 to 1979–80) was not outstanding, but moderate. All Indian growth for the agricultural output and yield during this period was recorded as 1.88% and 1.48% per annum, respectively. However, during this period few states achieved outstanding growth in the level of output and yield rate. In case of the level of output, the states like Manipur (6.46%), Arunachal Pradesh (5.9%), Maharashtra (5.78%), Punjab (5.43%) and Haryana (3.36%) performed significantly, while in case of yield the states Punjab, Maharashtra and Manipur achieved significant growth. In fact, during this period the effect of green revolution did not spread all over the country, which is visible from the observed lower growth rate of other states like Andhra Pradesh, Madhya Pradesh, Kerala, Rajasthan, Orissa, Bihar, Karnataka and Jammu and Kashmir in both output and yield. The effect of the green revolution was visible from the second phase of green revolution, i.e. the period 1980–81 to 1990–91. It has been observed from the kinked exponential growth rate (see Tables 1.1 and 1.2 below) that all Indian agricultural performance registered an unprecedented improvement in both output and yield growth during the second phase of the green revolution. The kinked exponential growth rate

[1] $\text{In}Y_t = a + b_t + u_t$ (Semi logarithmic equation is estimated to derive the exponential growth rate).

Table 1.1. Trend Growth Rate of Value of Output of Agriculture of Indian States 1970–71 to 2005–06

Growth rate	Exponential Growth rate		Kinked exponential growth rate	
	Whole period	1st sub-period	2nd sub-period	3rd sub-period
	(1970–71 to 2005 –06)	1970–71 to 1979–80	1980–81 to 1990–91	1990–91 to 2005–06
			b1	b2
Andhra Pradesh	2.406	1.93	3.16	1.46
Arunachal Pradesh	1.065	5.9	3.87	2.32
Assam	2.098	1.67	2.63	1.45
Bihar	2.549	0.61	2.26	1.98
Goa	3.3	3.01	1.46	3.12
Gujarat	2.082	2.76	-0.37	3.29
Haryana	3.463	3.36	4.78	2.34
Himachal Pradesh	2.797	1.4	2.63	1.08
Jammu & Kashmir	-0.263	1.79	1.75	-0.03
Karnataka	2.251	2.1	3.94	2.07
Kerala	1.189	-0.8	2.76	2.57
Madhya Pradesh	2.76	-0.93	4.61	1.84
Maharashtra	3.38	5.78	2.2	3.94
Manipur	2.8	6.46	0.45	2.16
Orissa	1.57	0.44	1.66	0.17
Punjab	2.88	5.43	4.9	1.47
Rajasthan	3.36	0.83	5.5	2.58
Tamil Nadu	1.896	1.77	5.46	0.15
Uttar Pradesh	1.918	1.09	3.14	1.58
West Bengal	3.24	2.26	5.25	2.09
Delhi	2.054	1.59	-1.9	1.23
India	2.6	1.88	3.34	2

Source: CSO, different issues, Govt of India Publication.

Note: The estimates are based on author's own calculation.

Table 1.2. Trend growth rate of yield from agriculture of Indian states 1970–71 to 2005–06

States	Exponential Growth Rate		Kinked Exponential Growth Rate	
	Whole period	1st sub-period	2nd sub-period	3rd sub-period
	1970 –2006	1970–80	1980–91	1991–92 to 2008–09
Andhra Pradesh	1.87	2.56	2.64	1.56
Assam	0.4	-0.37	2.98	0.021
Bihar	1.91	-0.0005	3.86	1.24
Gujarat	1.35	2.26	-0.34	2.93
Haryana	2.5	2.5	6.25	0.19
Himachal Pradesh	0.27	1.053	4.46	-0.34
Jammu & Kashmir	-2.79	0.69	2.08	-1.22
Karnataka	0.82	2.5	2.73	2.06
Kerala	0.7	-0.62	3.42	1.58
Madhya Pradesh	2.81	-1.21	6.1	1.95
Maharashtra	1.92	4.54	2.32	3.02
Manipur	1.64	3.23	3.21	-0.41
Orissa	-0.64	-0.19	2.22	-0.28
Punjab	2.12	3.9	6.38	-1.07
Rajasthan	2.52	0.56	4.76	1.79
Tamil Nadu	1.96	3.15	4.33	1.53
Uttar Pradesh	2.04	0.73	4.71	0.19
West Bengal	0.78	1.25	5.48	-0.29
India	1.43	1.48	3.38	1.16

Source: CSO, different issues, Govt. of India Publication.
Note: The estimates are based on author's own calculation.

of value of output and yield for India were 3.34% and 3.38% per annum, respectively, during this period. Most of the states grew at an accelerated rate in both output and yield during this period, and Tamil Nadu (5.46%), West Bengal[3] (5.25%), Rajasthan (5.5%), Punjab (4.9%), Haryana (4.78%) and Madhya Pradesh (4.62%) achieved outstanding growth in agricultural output, and Punjab (6.38), Haryana (6.29), Madhya Pradesh (6.11) and West Bengal (5.48) achieved remarkable growth in yield rate.

[3] See Ghosh (2010a and 2010b).

Moreover, during this period the states of different regions i.e. West Bengal of the eastern region, Rajasthan of the central region, and Tamil Nadu of the southern region contributed to the higher growth rate in India. All the other states registered higher growth during this phase compared to the previous period, barring a few exceptions.

However, it has been observed from this study that the phase of outstanding growth could not be sustained until the reform period, and the agricultural output and yield growth of the country experienced a severe slowdown during the post reform period. All Indian growth in output and yield rates declined to 2% and 1.16% per annum during this period, compared to 3.34% and 3.38% in the previous period. Aside from Maharashtra, Gujarat and Goa, all other states experienced a significant deceleration in both output and yield growth in the reform period. Some states also registered negative growth during this period. In fact, according to many researchers the most important factor for this agricultural downturn may be traced in the slowdown of food grain production in the post-reform period in India. Since foodgrain comprises nearly 60% of the total crop output in India and the majority of states are dependent on its production, the slackening growth of value of crop output reflects the deceleration in food grain production. Thus, the results establish that there was a significant acceleration in growth rate in the value of output and yield during the second phase of the green revolution, but the post liberalization period has marked a sign of depression both in agricultural output and its yield rate in India. Our findings are in conformity with the previous empirical studies by Bhalla & Singh, and others. In a comprehensive study about the growth and regional disparity of Indian states in respect of agricultural output, yield and area, Bhalla & Singh (2009) argued that increase in regional disparity during the first phase of the green revolution (1971–81) was followed by decline during the second phase due to the effect of HYV technology. However, the post liberalization period (since 1991) witnessed the continuous decline in growth rates in Indian agriculture. This result was supported by authors like Mathur etal. (2006), Mahendradev (1987), Janaiah et al.(2005) and Chand et al. (2007),according to whom factors like the slowdown in growth of fertilizer use, declining irrigation intensity, reduction of energy use and lack of adoption of modern techniques were considered to be attributable to the deceleration in the trends of agricultural growth in India during the 1990s.

Table 1.3. Change in percentage of area under foodgrain and non-foodgrain crops of Indian states

States	FOODGRAIN			NON-FOODGRAIN		
	1970–71	1990–91	2007–08	1970–71	1990–91	2007–08
Andhra Pradesh	71.07	56.33	54.45	3.3	7.15	10.18
Assam	75.44	71.24	65.59	11.6	13.07	10.32
Bihar	89.86	80.24	88.86	4.4	6.46	6.05
Gujarat	50.78	37.10	36.66	16.5	15.81	24.33
Haryana	77.58	64.36	69.31	9.7	25.11	19.21
Himachal Pradesh	90.15	87.42	83.67	2.7	2.56	2.58
Jammu & Kashmir	89.41	83.76	81.01	3.1	5.55	5.01
Karnataka	66.00	58.01	61.05	10.2	7.26	6.00
Kerala	31.58	19.01	8.80	0.5	0.55	0.12
Madhya Pradesh	81.96	66.12	55.29	4.7	6.83	6.52
Maharashtra	66.93	63.83	64.69	15.8	15.78	21.13
Orissa	68.50	73.90	60.88	1.8	2.76	1.11
Punjab	69.16	75.10	80.09	11.4	12.10	10.43
Rajasthan	76.96	62.39	61.27	3.1	16.03	13.00
Tamil Nadu	70.36	57.43	53.27	6.2	7.18	9.61
Uttar Pradesh	84.59	78.71	76.56	16.1	12.58	14.24
West Bengal	86.22	73.80	65.17	8.7	14.40	14.77
India	74.26	66.87	63.35	8.8	10.96	11.57

Source: Data for foodgrain and non-foodgrain crops are compiled from different issues of CMIE, Agriculture. Foodgrain crops: rice and wheat, Non-foodgrain crops: jute, cotton, sugarcane, rapeseed, mustard and potato.
Note: Estimates are based on Author's own calculations.

3.2 Cropping pattern in India

Cropping pattern is defined by the proportion of area under different crops at a point of time. It is a mix of agricultural crops grown in a particular geographical area. Changes in cropping pattern can be seen as a change in the proportion of area or value of production under different crops to total agricultural area over time. The cropping pattern is governed by the law of comparative advantage in relation to agro climatic conditions, and technical and institutional factors (Vaidyanathan 1994). In India, since the 1980s the process of change in the cropping pattern hastened and was prominent in favour of non-foodgrain crops, reflected in the ratio of area in foodgrain crop to GCA. During the period 1970–71 to 1980–81 the proportion of area under foodgrain experienced a small decline from 74.3% in 1970–71 to 72.6% in 1980–81 (see Table 1.3). Since the 1980s onwards, the share of area under foodgrain to total GCA declined persistently to 66.9% in 1991–92, 64.6% in 2000–01 and 63.4% in 2007–08, respectively. Taking all selected non-foodgrain crops together, the area allocation reveals an upward trend at the all-Indian level over the period 1970–71 (8.8%) to 2007–08 (11.6%).

3.3 State wise changes in cropping patterns

State wise patterns of changes in area of crops overtime in India are wide and varied. It is worth mentioning that foodgrain occupies the most important place in area composition in almost all states still now. India is mostly a foodgrain-producing country. In recent years, especially in the post-reform period, a marginal shift in cropping patterns favouring non-food crops is observable. However, there are substantial inter-state variations to this result. In the case of foodgrain, the proportion of area declined at a high rate in states like Andhra Pradesh, Gujarat, Rajasthan, Tamil Nadu, Madhya Pradesh and West Bengal. Haryana, Himachal Pradesh, Jammu & Kashmir and Uttar Pradesh also experienced deceleration in the composition of area under foodgrain, but at a moderate rate. In Bihar around 90% of GCA was under foodgrain crop in 1970–71. In 1980–81 this decreased to 66.55%. Again, the same proportion registered a sharp increase from 66.55% to 80.2% in 1991–92, 90.7% in 2000–01 and 88.9% in 2007–08, respectively. Further, the state of Punjab experienced an increase in the proportion of area under foodgrain from 69.2% in 1970–71 to 80.1% in 2007–08. Excepting Punjab and Bihar, almost all the states showed a diversification of cropping pattern in favour of non-food crops.

Andhra Pradesh, Gujarat, Haryana, Jammu & Kashmir, Madhya Pradesh, Maharashtra, Rajasthan, Tamil Nadu and West Bengal experienced an increase in the area allocated to major non-food crops over the study period. Among the 17 states considered here, more than 9 faced a deceleration in the share of non-foodgrain crops after the year 1991–92.

Among the remaining states, Andhra Pradesh, Maharashtra and Tamil Nadu managed to maintain a steady upward trend in terms of area allocation towards selected non-foodgrain crops. Gujarat, Madhya Pradesh, Punjab and Rajasthan experienced an increase in the area allocated to non-food crops, especially in the post-reform period. The crop as well as state-wise scenario is presented in appendix Tables A1 to A7.

3.4 Regional disparity in agricultural infrastructure in India: an interstate analysis

India is characterized by wide regional variation in agro-climatic condition. Agricultural output in different regions varies due to different agro-climatic factors, physical resource endowment and level of investment in rural infrastructure and technological innovation.[4] The regional variation in agricultural infrastructure and the use of agricultural inputs in India is quite high.

To provide a vivid picture[5] a composite index of agricultural infrastructure, constructed through the Deprivation Method, explores the disparity in agricultural infrastructure across the states of India. The detailed methodology of constructing the index is explained in the methodology section. The result of the index and the subsequent rank of the states according to the index are presented in Table 1.4. From the table it is clear that there is a widespread disparity among states in respect of the distribution of agricultural inputs in India. The advanced states consistently enjoyed the benefit of better agricultural infrastructure throughout the period under study. For example, states like Punjab, Haryana, Uttar Pradesh, Tamil Nadu and Andhra Pradesh occupied the top positions according to the rank of the index, whereas Assam, Himachal Pradesh and Orissa lagged behind throughout the period. Assam and Orissa remained at the bottom throughout the study period. Kerala, Karnataka, Gujarat, Maharashtra, Rajasthan and West Bengal occupied middle positions during the periods under study. The status of Jammu and Kashmir declined

---

[4] Apart from these physical factors, agrarian relations and tenurial contracts play an important role in enhancing agricultural productivity and efficiency. For details see Laha & Kuri (2012).

[5] To construct this index eight agricultural development indicators are selected: Cropping intensity (CI), Percentage irrigated area to GCA (IAGC), Fertilizer consumption per hectare of GCA (fcgc), Credit to agriculture (CTA), Number of tractors and pumpsets used per 1,000 hectares (TAP), Average yield of agricultural land (AY), Road length per 100 sq km (RL), Percentage share to total consumption of electricity in agriculture (CELA).

Table 1.4. Composite index of agriculture infrastructure of Indian states 1980–81 to 2007–08 (CIAI)

STATES	INDEX				RANK			
	1980–81	1990–91	2000–01	2007–08	1980–81	1990–91	2000–01	2007–08
Andhra Pradesh	0.33	0.52	0.57	0.58	5	4	4.5	4
Assam	0.10	0.10	0.10	0.08	16.5	17	17	16
Bihar	0.21	0.29	0.39	0.39	11	8.5	8	8
Gujarat	0.28	0.31	0.32	0.45	7	7	10	6
Haryana	0.60	0.72	0.73	0.76	2	2	2	2
Himachal Pradesh	0.17	0.17	0.16	0.16	13.5	15	15	15
Jammu & Kashmir	0.22	0.22	0.20	0.23	9	12.5	14	14
Karnataka	0.17	0.33	0.41	0.37	13.5	6	7	9
Kerala	0.31	0.28	0.33	0.27	6	10	9	13
Madhya Pradesh	0.10	0.21	0.22	0.31	16.5	14	13	12
Maharashtra	0.21	0.26	0.28	0.32	11	11	11	11
Orissa	0.12	0.13	0.12	0.07	15	16	16	17
Punjab	0.90	0.94	0.85	0.85	1	1	1	1
Rajasthan	0.24	0.22	0.27	0.34	8	12.5	12	10
Tamil Nadu	0.51	0.50	0.57	0.54	4	5	4.5	5
Uttar Pradesh	0.53	0.61	0.61	0.63	3	3	3	3
West Bengal	0.21	0.29	0.42	0.41	11	8.5	6	7

Source: Data for inputs are taken from various issues of Indian Agriculture in brief and CMIE, Agriculture.
Notes: Estimates are based on the author's calculations.

to a large extent as its position slipped from ninth in 1980–81 to fourteenth in 2007–08 during the period of study. The reverse happened for Madhya Pradesh as its position improved from 16.5 in 1980–81 to 12 in 2007–08. Thus, from the result of composite index it can be said that a wide disparity remained in the distribution of agricultural infrastructure across the states in India during the study period.

3.5 Agricultural growth in India: a convergence analysis

Several attempts have been made in India to explain the nature of convergence in agricultural growth. Kalirajan et al. (1998) found regional divergence in agricultural production and productivity in India. According to them, the investment climate in agriculture and operation of both the supply and demand factors are responsible for slowing growth rates and the regional divergence in agricultural production and productivity. Similarly, Ghosh (2006) found absolute β and σ divergence in land productivity, labour productivity and per capita agricultural output across the states after the dissemination of new HYV technology and large-scale economic reforms. He also found conditional convergence in agricultural development taking human capital, physical capital and rural infrastructure as conditional variables. Against this backdrop, sigma and absolute β convergence have been tested over the period 1970–71 to 2007–08 for major Indian states to analyze the nature of convergence/divergence in agricultural output.

Sigma (σ) convergence is tested by computing the standard deviation of logarithm of per capita value of agricultural output across regions. The result is displayed in Fig. 1.1. The standard deviation increased from 0.4 in 1970–71 to 0.53 in 2008–09. This indicates a trend of divergence. Fig. 1.1 shows that during 1975–76 disparity increased considerably but slowed during the 1980s, i.e. in the second phase of the green revolution, but since 1991–92 it again picked up and continued to rise until 2008–09. Therefore, the test of sigma convergence confirms the divergence of per capita value of agricultural output across the Indian states over time.

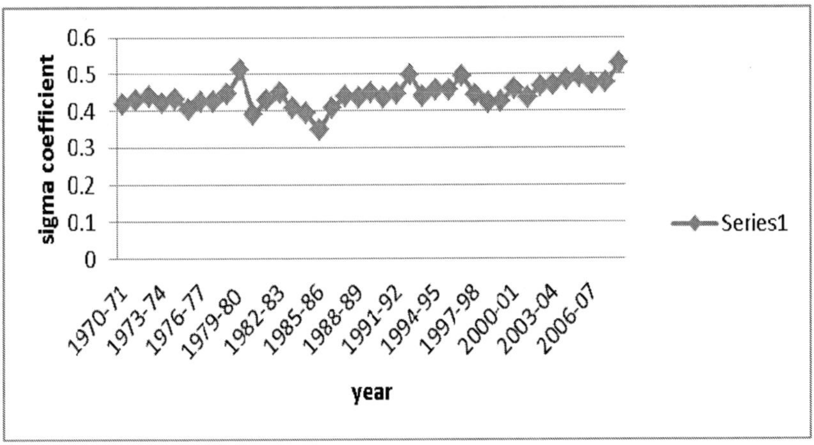

Fig. 1.1. Variation in sigma coefficient of per capita value of agricultural output

Absolute β convergence is tested in the second step to reaffirm the divergence in PCVOA. β convergence is tested by regressing the growth rate of PCVOA on the initial level of PCVOA. The regression equation used for this analysis is as follows:

$$Gi,t,t\text{-}\tau = [\ln(Yi,t) - \ln(Yi,t\text{-}\tau)]/\tau = \alpha + \beta\ln(Yi,t\text{-}\tau) + \varepsilon i,t \text{ -(1)}$$

$Gi,t,t\text{-}\tau = [\ln(Yi,t) - \ln(Yi,t\text{-}\tau)]/\tau$ = ith region's average growth rate of PCVOA between the period t and t-τ.ln(Yi,t) and ln(Yi, t-τ) are the natural logarithms of ith region's PCVOA at time t and t-τ respectively. τ indicates the length of period. Negative sign of regression coefficient implies absolute β convergence, otherwise divergence (Ghosh 2006). In this analysis, four regressions have been fitted-one for the whole period, another two for the pre and post-reform periods, and the last one for the last decade, i.e. for the period 2000–01 to 2008–09. The results of absolute β convergence are depicted in Table 1.5.

Table 1.5: Regression result of absolute β convergence of PCVOA

Period	β coefficient	t value	significance	R^2
1970–71 to 2008–09	9.23E-03	1.146	0.266	0.065
Pre Reform Period				
1970–71 to 1990–91	1.40E-02	1.14	0.268	0.064
Post Reform Period				
1991–92 to 2008–09	5.68E-03	2.535	0.02	0.254
2000–01 to 2008–09	6.27E-03	0.776	0.447	0.031

Source: Estimates are based on author's calculations

The results show a positive β coefficient for the whole period and the sub-periods. This implies absolute β divergence in PCVOA over all periods. Thus, the results of both σ and β convergence analyses indicate a sign of divergence in agricultural performance across the states in the country.

3.6 Unit root test of divergence: an interstate analysis

Historically, the conventional cross sectional regression for determining convergence has come under criticism. Quah (1993) showed that this bias is similar to Galton's Fallacy, whereas Friedman (1992) comments that convergence is indicated by a diminution of the variance among countries overtime. Several studies have relied on time series information for determining the existence, or lack thereof, of convergence on a cross section study (Ben-David 1993; Bernard & Duarlof 1996; Li & Papell 1999; Cheung & Pascual 2004). The long run forecast of difference in PCGDP between any pair of countries converges to zero as the forecast horizon grows according to this new methodology. Within a neoclassical set up the test for convergence of per capita income is translated to a test for the stationarity of output differential. In this context we have examined the behaviour of each state's per capita agricultural output differential with the national average overtime to ascertain whether there is any noticeable evidence of a trend of convergence. The statistical tests of long-run convergence hinge on the time series properties of $[In(Y_{i,t})\text{-}In(\text{-}Y_{*t})]$,where $In(Y_{i,t})$ and $In(\text{-}Y_{*t})$ are the logarithm of per capita value of output of ith state and the national average, respectively. In this section the behaviour of $[In(Y_{i,t})\text{-}In(\text{-}Y_{*t})]$ has been tested for the period 1980–81 to 2008–09 for 19 Indian states.

Table 1.6: Philips-Perron unit root test for convergence of PCVOA of Indian states

State	PP Test Statistics	P-value
Andhra Pradesh	-4.6*	0
Arunachal Pradesh	-2.84	0.19
Assam	-5.27*	0
Bihar	-3**	0.14
Goa	-5.78*	0
Gujarat	-4.68*	0
Haryana	-3.15	0.01
Himachal Pradesh	-4.28*	0
Jammu & Kashmir	-2.79	0.2
Karnataka	-3.7**	0.03
Kerala	-2.42	0.36
Madhya Pradesh	-1.82	0.36
Maharashtra	-4.18**	0.01
Orissa	-4.89*	0.002
Punjab	-3.16	0.01
Rajasthan	-6.37*	0
Tamil Nadu	-3.48**	0.01
Uttar Pradesh	-4.19**	0.01
West Bengal	-2.56	0.29

Notes:* and ** denote significance at 1% and 5% level, respectively.
Source: author's calculations

The results of the unit root test for convergence show a diverging pattern across the states in the case of agricultural output, shown in Table 1.6. The null hypothesis of unit root for the Phillips Peron test is rejected for the states of Andhra Pradesh, Assam, Goa, Himachal Pradesh, Karnataka, Maharashtra, Gujarat, Rajasthan, Orissa, Tamil Nadu and Uttar Pradesh, whereas the existence of unit root is accepted for the states of Arunachal Pradesh, Bihar, Jammu& Kashmir, Kerala, Madhya Pradesh, Punjab, Haryana and West Bengal. Therefore, while 11 states are converging towards national average in PCVOA overtime, the eight states are following a different and steady state from the national average. Interestingly, out of eight states, Punjab, Haryana and West Bengal belong to the category of agriculturally advanced states while the remaining five of Arunachal Pradesh, Bihar, Jammu& Kashmir, Kerala and Madhya Pradesh belong to the category of backward states. Thus, our results establish the argument that the wide variety in agricultural infrastructure among these eight states is responsible for wide variability in agricultural production, thus making the growth path divergent.

4. Conclusion

This study examines the performance of major Indian states at the level of agricultural development and also the disparity prevailing in terms of agricultural performance. The estimated exponential and kinked exponential growth rates of value of agricultural output and yield for three periods reveal that the outstanding and remarkable growth observed by Indian states during the 1980s deteriorated significantly during the 1990s. The cropping pattern in India is mainly biased towards foodgrain crops. However, a change in cropping pattern is observed in favour of some non-foodgrain crops during the study period. This study reveals that, except for Punjab and Bihar, the area allocated towards foodgrain crops decreased for all the states during the study period. In the case of non-foodgrain crops, Andhra Pradesh, Gujarat, Haryana, Jammu& Kashmir, Madhya Pradesh, Maharashtra, Rajasthan, Tamil Nadu and West Bengal experienced an increasing trend of crop diversification. Moreover, the construction of CIAI shows a high degree of regional variation in agricultural infrastructure across states. The ranking structure of CIAI indicates that the states with developed infrastructure performed much better in terms of growth of agricultural output and productivity in all the periods. The convergence analysis depicts the evidence of both σ divergence and absolute β divergence in PCVOA for the period 1980–81 to 2007–08. Moreover, the unit root test establishes that out of 19 states, 11 are converging towards the national average in PCVOA overtime, while the remaining eight are following a different and steady state from national average. The wide variety in agricultural infrastructure among these eight states is responsible for the wide variability in agricultural production in India.

Appendix

Table AI. Change in the rice cropping pattern of Indian states 1970–71 to 2007–08

States	RICE				
	1970–71	1980–81	1990–91	2000–01	2007–08
Andhra Pradesh	26.4	29.4	29.8	31.3	29.4
Assam	71.0	66.0	65.5	65.4	60.5
Bihar	47.8	34.3	45.8	45.9	45.2
Gujarat	4.9	5.3	5.7	5.6	6.2
Haryana	5.4	8.9	11.5	17.2	16.6
Himachal Pradesh	10.5	11.6	8.5	8.6	8.1
Jammu &Kashmir	25.6	27.2	25.2	21.9	23.2
Karnataka	10.7	10.5	10.2	12.1	11.0
Kerala	29.8	28.0	17.9	11.5	8.3
Madhya Pradesh	21.3	7.1	8.7	9.4	7.6
Maharashtra	7.0	7.2	7.7	5.6	6.9
Orissa	53.4	47.9	46.3	56.3	49.4
Punjab	6.9	17.5	27.6	32.9	33.2
Rajasthan	0.7	1.0	0.8	0.9	0.6
Tamil Nadu	36.4	35.5	30.4	33.3	30.8
Uttar Pradesh	19.7	20.0	21.4	23.1	22.9
West Bengal	69.9	67.9	66.9	59.6	58.7
India	22.5	22.9	23.4	23.9	22.4

Source: author's calculation

Table A2. Change in the wheat cropping pattern of Indian states 1970–71 To 2007–08

WHEAT					
States	**1970–71**	**1980–81**	**1990–91**	**2000–01**	**2007–08**
Andhra Pradesh	0.1	0.1	0.1	0.1	0.1
Assam	0.7	3.0	2.0	1.7	1.5
Bihar	11.9	15.1	23.9	26.4	27.3
Gujarat	5.7	5.8	3.9	2.7	10.4
Haryana	22.8	27.0	32.5	38.4	38.1
Himachal Pradesh	33.2	37.0	38.5	39.9	37.8
Jammu & Kashmir	21.2	20.3	22.9	24.8	24.5
Karnataka	2.8	2.9	1.6	2.2	2.1
Madhya Pradesh	16.5	15.2	19.8	15.0	18.3
Maharashtra	4.6	5.3	3.1	3.5	5.5
Orissa	0.2	0.8	0.3	0.1	0.1
Punjab	40.5	41.6	43.1	42.9	44.3
Rajasthan	8.8	9.4	9.8	12.0	11.7
Uttar Pradesh	25.5	31.1	34.3	36.2	36.6
West Bengal	5.1	3.7	2.9	4.7	3.6
India	10.9	12.9	12.8	13.5	14.3

Source: author's calculation

Table A3. Change in jute cropping pattern of Indian states 1970–71 to 2007–08

JUTE					
State	**1970–71**	**1980–81**	**1990–91**	**2000–01**	**2007–08**
Assam	4.7	3.3	2.6	1.7	1.6
Bihar	1.3	1.4	1.9	1.7	1.7
Orissa	0.5	0.5	0.4	0.0	0.1
West Bengal	5.7	8.0	6.7	6.7	6.3
India	10.9	12.9	12.8	13.5	14.3

Source: author's calculation

Table A4. Change in cotton cropping pattern of Indian states 1970–71 to 2007–08

Cotton					
States	**1970–71**	**1980–81**	**1990–91**	**2000–01**	**2007–08**
Andhra Pradesh	2.4	3.4	5.4	7.5	8.4
Gujarat	15.7	14.6	10.8	15.5	19.3
Haryana	3.9	5.8	10.6	9.1	7.5
Karnataka	9.1	9.5	4.7	4.5	3.1
Kerala	0.3	0.2	0.4	0.1	0.0
Madhya Pradesh	3.4	2.8	3.2	2.6	3.1
Maharashtra	14.6	12.6	13.5	14.2	15.6
Orissa	0.0	0.0	0.1	0.5	0.6
Punjab	7.0	9.6	8.8	6.0	7.7
Rajasthan	1.3	2.1	2.6	2.7	1.7
Tamil Nadu	4.2	3.4	3.7	2.7	3.5
Uttar Pradesh	0.2	0.2	0.1	0.0	0.0
India	4.5	4.5	4.2	4.6	4.8

Source: author's calculation

Table A5. Change in sugarcane cropping pattern of Indian states 1970–71 to 2007–08

SUGARCANE					
States	**1970 –71**	**1980 –81**	**1990 –91**	**2000 –01**	**2007 –08**
Andhra Pradesh	0.9	1.1	1.8	1.8	1.8
Assam	1.2	1.4	1.0	0.7	0.7
Bihar	1.5	0.9	1.8	1.2	1.4
Gujarat	0.4	0.9	1.1	1.7	1.7
Haryana	3.1	2.1	2.9	2.3	2.2
Himachal Pradesh	0.4	0.3	0.2	0.3	0.3
Jammu & Kashmir	0.4	0.3	0.2	0.3	0.3
Karnataka	0.9	1.4	2.3	3.4	2.4
Kerala	0.3	0.3	0.3	0.1	0.1
Madhya Pradesh	0.3	0.3	0.3	0.3	0.4
Maharashtra	1.1	1.3	2.2	2.8	5.4
Orissa	0.4	0.8	0.4	1.3	0.2
Punjab	2.3	0.7	1.4	0.3	0.3
Rajasthan	0.2	0.2	0.2	0.1	0.0
Tamil Nadu	1.8	2.8	3.4	5.0	6.1
Uttar Pradesh	5.8	5.5	7.6	7.7	8.7
West Bengal	0.5	0.2	0.2	0.2	0.2
All India	1.6	1.5	2.1	2.3	2.6

Source: author's calculation

Table A6. Change in rapeseed and mustard cropping pattern of Indian states 1970–71 to 2007–08

RAPESEED & MUSTARD					
States	**1970–71**	**1980–81**	**1990–91**	**2000–01**	**2007–08**
Assam	4.9	6.2	7.9	6.7	6.1
Bihar	0.8	0.6	1.1	1.3	1.1
Gujarat	0.3	1.6	3.7	1.8	2.8
Haryana	2.6	5.5	11.5	6.6	9.3
Himachal Pradesh	0.4	0.6	0.9	1.1	0.9
Jammu & Kashmir	2.9	4.1	5.5	5.7	5.0
Karnataka	0.1	0.0	0.0	0.1	0.0
Madhya Pradesh	1.0	0.8	3.2	2.5	2.8
Maharashtra	0.0	0.0	0.0	0.0	0.0
Orissa	0.7	1.9	1.8	0.2	0.2
Punjab	1.8	1.9	1.5	0.7	0.4
Rajasthan	1.5	2.1	13.2	8.0	11.2
Uttar Pradesh	9.3	1.7	3.5	3.6	3.5
West Bengal	1.5	1.7	4.8	4.8	4.2
All India	2.0	2.3	3.6	2.4	3.0

Source: author's calculation

Table A7. Change in potato cropping pattern of Indian states 1970–71 to 2007–08

POTATO					
States	**1970–71**	**1980–81**	**1990–91**	**2000–01**	**2007–08**
Assam	0.9	1.1	1.6	2.0	2.0
Bihar	0.9	0.9	1.7	1.9	1.9
Gujarat	0.0	0.1	0.2	0.3	0.6
Haryana	0.1	0.2	0.2	0.3	0.3
Himachal Pradesh	1.9	1.4	1.5	1.4	1.4
Karnataka	0.1	0.1	0.2	0.3	0.5
Madhya Pradesh	0.1	0.1	0.2	0.2	0.2
Maharashtra	0.1	0.1	0.1	0.1	0.1
Orissa	0.2	0.1	0.1	0.1	0.1
Punjab	0.3	0.6	0.4	0.8	1.0
Rajasthan	0.0	0.0	0.0	0.0	0.0
Tamil Nadu	0.2	0.2	0.1	0.1	0.1
Uttar Pradesh	0.7	1.1	1.5	1.6	2.0
West Bengal	0.9	1.5	2.7	3.3	4.1
All India	0.3	0.4	0.6	0.7	0.8

Source: author's calculation

References

Arellano, M., & S. Bond. "Some Tests of Specification for Panel Data: Monte Carlo Evidence and an Application to Employment Equations." *Review of Economic Studies* 58 (1991): 277–97.

Barro, R. J. & X. Sala-i-Maritn. "Convergence." *Journal of Political Economy* 100 (1992): 223–251.

—. "Convergence across States and Regions." *Brookings Papers on Economic Activity* I (1991): 107–18.

—. *Economic Growth.* McGraw-Hill: New York, 1995.

Ben-David, Dan. "Equalizing Exchange: Trade Liberalization and Income Convergence." *Quarterly Journal of Economics* 108 (3) (1993): 653–79.

Bernard, A. & S. Duarlauf. "Interpreting Tests of the Convergence Hypothesis." *Journal of Econometrics* 10 (1996): 97–108.

Bhalla, G. S. & G. Singh. "Economic liberalisation and Indian agriculture: A statewise analysis." *Econ. Political Weekly* 44 (52) (2009): 34–44.

Boyce, J. K. *Agrarian Impasse in Bengal: Institutional Constraints to Technical Change.*Oxford: Oxford University Press, 1987.

Central Statistical Organisation. *Statewise and Cropwise Estimates of Value of Output from agriculture in India during 1960–61 to 1980–81.* National Account Statistics, Ministry of Statistics and Programme Implementation, Government of India, 1985.

—. *Statewise and Cropwise Estimates of Value of Output from agriculture in India during 1980–81 to 1990–91.* National Account Statistics, Ministry of Statistics and Programme Implementation, 1996.

—. *Statewise and Cropwise Estimates of Value of output in agriculture in India during 1990–91 to 2002–03.* National Account Statistics, Ministry of Statistics and Programme Implementation, 2008.

Centre for Monitoring Indian Economy (CMIE). "Agriculture." Mumbai, 2010.

Chand, R., S. S. Raju & L. M. Pandey. "Growth crisis in agriculture: Severity and options at national and state level." *Economic and Political Weekly* 42 (26) (2007): 2528–2533.

Cheung, Y. W. & A. G. Pascual. "Testing for Output Convergence: A Re-examination." *Oxford Economic Papers* 56 (1) (2004): 45–63.

Desai, G. M. & P. B. R. Hazell. "Instability in Indian Foodgrain Production." *International Food Policy Research Institute* 30 (1982).

EPW Research Foundation. "Domestic Product of States of India: 1960–61 to 2006–07." 2009.

Friedman, M. J. "Do old Fallacies Ever Die?" *Journal of Economic Literature* 30 (4) (1992): 2129–32.

Ghosh, B. K. "Fading Glory of West Bengal Agriculture in the context of globalization: Need for a change in Cropping Pattern." *Trends in Agricultural Economics* 3 (4) (2010): 185–189.

—. "Growth and Variability in the Production of Crops in West Bengal Agriculture." *Trends in Agricultural Economics* 3 (2010): 135–146.

Ghosh, M. "Regional convergence in Indian agriculture." *Indian J. Agric. Eco.* 61 (2006): 610–629.

Greene, W. H. *Econometric Analysis.* Vol. Fifth Edition. Pearson Education (Singapore) Pvt. Ltd., Patparganj, Delhi, India., 2003.

Janaiah, A., K. Otuska, & M. Hossain. "Is the productivity impact of the green revolution in rice vanishing? empirical evidence from TFP analysis." *Econ. Political Weekly* 40 (2005): 5596–5600.

Kalirajan, K. P., S. Bhid & R. T. Shand. "India's agricultural dynamics: Weak link in development." *Econ. Political Weekly* 33 (39) (1998).

Laha, A. & P. K. Kuri. "Differences in resource allocation under alternative tenurial contracts and its explanations: Evidence from rural West Bengal." *Trends in Agricultural Economics* (2012): 13–22.

Li, Q. & D. Papell. "Convergence of International Output: Time Series Evidence for 16 OECD Countries." *International Review of Economics and Finance* 8 (1999): 267–280.

Mahendradev, S. "Growth and instability in foodgrains production: An inter-state analysis." *Econ. Political Weekly* 22 (1987).

Mathur, A. S., S. Das & S. Sircar. "Status of agriculture in India: Trends and prospects." *Econ. Political Weekly* 41 (2006): 5327–5336.

Mehra, S. "Instability in Indian Agriculture in the Context of new Technology." *International Food Policy Research Institute, India*, 1981.

Quah, Danny T. "Galton's Fallacy and Tests for Convergence Hypotheses." *The Scandinavian Journal of Economics* 95 (4) (1993): 427–443.

Rao, C. H. H. *Technological Change and the Distribution of Gains in Indian Agriculture.* New Delhi: MacMillan, India, 1975.

Reserve Bank of India. "Handbook of Statistics on State Government Finances." 2010.

Sawant, S. D. & C. V. Achuthan. "Agricultural growth across crops and regions: Emerging trends and patterns." *Econ. Political Weekly* 95 (4) (1995).

Tiffin, R. & X. Irz. "Is agriculture the engine of growth?" *Agricultural Economics* 35(1) (2006): 79–89.

Yang, D. T. & X. Zhu. "Modernization of agriculture and long-run growth." Preliminary and Incomplete Notes, 2004.

CHAPTER TWO

ASSESSMENT OF THE ECOLOGICAL FOOTPRINT OF AGRICULTURAL LAND USE OF WEST BENGAL, INDIA

BISWAJIT GHOSH AND NAMITA CHAKMA

1. Introduction

An ecological footprint represents the human impact on the earth in a clear manner (Moffatt 2000), originally conceived as a simple and elegant method for comparing the sustainability of resource use among different populations (Rees 1992). The concept was further developed by Wackernagel & Rees (1995) and Wackernagel et al. in 1999. According to Costanza (2000), it is a method to highlight the unsustainability of global consumption. Ecological footprint is a measurement of the amount of land area required to sustain a population of any given size. Under prevailing technology, it measures the amount of arable land and aquatic resources that must be used to continuously sustain a population, based on its consumption levels at a given point in time. To the fullest extent possible, this measurement incorporates water and energy use, usage of land for infrastructure and different forms of agriculture, forests and all other forms of energy and material inputs that people require for their day-to-day living. It also accounts for the land area required for waste assimilation (Li &Li 2009).

Recent years have seen the extensive application of the concept of ecological footprint recent years (Muniz &Galindo 2005; Haberl et al. 2001; Stoglehner 2001; Kathryn 2000; Hanley et al. 1999). It is a relatively new concept in the study of geography, credited due to its simple method and simple representation of the world in terms of ecological sustainability.

However, a significant number of researchers have criticized the concept since its inception (Levett 1998; Moffatt 2000; Ayres 2000; Rapport 2000) over its simplification and difficulty in understanding any specific reason for unsustainability of consumption of a population, and that it disregards technological advancement and its impact. Therefore, its utility is time and space specific.

Irrespective of its limitations, one cannot simply ignore the crucial fact that we are now living in a world with limited land resources. With populations increasing at an alarming rate and alternative uses of croplands creating an ever widening gap between demand and supply of food products, the development of the concept and debate are still continuing (Lenzen & Murray 2003).

In the present study, ecological footprint of agricultural land use indicates the land required to sustain a population for its food requirements and that is intended to measure the level of regional sustainable development of West Bengal, since agriculture is its main economic base. In this chapter an effort has been made to measure that gap so that remedial measures could be taken by the appropriate authority to maintain a considerable level of regional sustainable development.

2. Study Area

West Bengal (21°38′ to 27°10′N and 85°50′ to 89°50′E), as the name implies, is the western part of a larger regional entity named Bengal. More precisely, it is the area left out from Bengal Province in undivided India, after Bangladesh was carved out of it (Bose 1979). The state is a part of the eastern zone of India. It is essentially a flat, featureless alluvial plain, a large portion of which is a part of the Ganges river delta. The vast alluvial plains spread from Jalpaiguri and Siliguri in the north to the Sundarban creeks and the Kanthi littoral in the south. Only 1% of it in the far north, in the Darjeeling hills, is mountainous which forms a part of the world's highest mountain range, the Himalayas. The plateau fringe, a rolling upland, with small isolated hills standing here and there and the Purulia triangle of upland along its western border, comprises about 6% of the total area (Bose 1979).

Fig. 2.1. Location of the study area

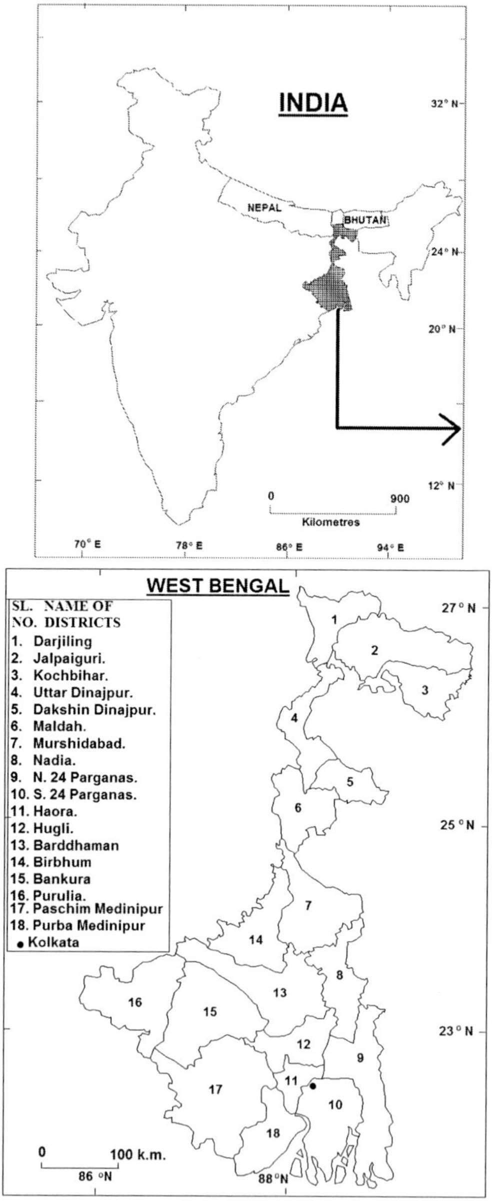

3. Agriculture as the main economic base of West Bengal: a brief account

Agriculture has traditionally been the backbone of West Bengal's economy. However, the amount of gross cropped area is 97,51,000 ha, the net sown area is 52,95,000 ha, and the cultivable area is 71,14,000 ha; 62% is irrigated, and the average land holding is 0.82 ha (Department of Agriculture, Government of West Bengal 2010). Successful cultivation presupposes suitable ecological and social conditions, and West Bengal provides suitable combination of these (Bose 1979).

West Bengal has an area of 88,752 km^2 (34,267 mi^2) consisting of nineteen districts. It has the share of 8% of India's population with 2.7% of its total area. Marginal (individual land holding 0-1 ha) and small (individual land holding 1-4 ha) farmers constitute 95% of the 5.7 million farmers, and there are 7.5 million farm workers. Agriculture contributes nearly 16 billion, or 24%, of the annual state domestic product. As per the provisional census of India, 2011, West Bengal is the fourth most populous state of India with an average population density of 1,029, which puts immense pressure on the land (see Table 2.1 and Fig. 2.2 below).

A great variety of crops is grown in the state, including Kharif (summer crops) and Rabi (winter crops) along with their intermediate varieties. The agricultural operation is conducted both on subsistence and commercial bases. Double cropping and multiple cropping are practised on suitable land in terms of irrigation facility. The state is the first in India in the production of rice, vegetables, inland fisheries and pineapple, second in the production of potato and tea, and third in the production of flowers (West Bengal Agricultural Highlights profile 2010).

Major cereal producing districts are Paschim Medinipur, Barddhaman, Purba Medinipur and Murshidabad. Pulses can be grown on a wide variety of soils that are not stiff clayey or very sandy. Murshidabad, Malda, Nadia, Birbhum and South 24 Parganas are the major pulse producing districts. Oilseed cultivation needs medium and high land with good drainage facilities and loamy and alluvium soil. Murshidabad, Nadia, Paschim Medinipur and North 24 Parganas are the major oilseed producing districts. The climatic condition of West Bengal is suitable for sugarcane. Murshidabad, Malda, Paschim Medinipur, Nadia, Barddhaman and Birbhum are important sugarcane producing districts. High temperatures, high rainfall and loamy to sandy soils are suitable for jute. Murshidabad, Nadia, North 24 Parganas, Hugli and Uttar Dinajpur are the major jute producing districts. Potato grows on sandy and loamy soil. The crop cannot withstand waterlogging. Hugli, Paschim Medinipur, Barddhaman,

Bankura and Jalpaiguri are the major potato producing districts. The production of tea is much higher in Jalpaiguri than in Darjeeling, but the quality of Darjeeling tea is world famous. A small quantity of tea is grown in the Kooch Bihar and Uttar Dinajpur districts as well. Tobacco grows well on the well-drained loamy and new alluvium soils. Kooch Bihar and Jalpaiguri are the main jute producing districts.

In the present work, eight important croplands of the state are selected for study. These are cereals, pulses, oilseeds, jute, potato, sugarcane, tea and tobacco.

Table 2.1. Population Density of West Bengal (2011)

SL.NO	STATE/DISTRICT	POPULATION	AREA IN SQ. KM	POPULATION DENSITY/SQ.KM.
WEST BENGAL		91,347,736	88,752	1029.24707
1	DARJEELING	1,842,034	3,149	584.9583995
2	JALPAIGURI	3,869,675	6,227	621.4348804
3	KOCH BIHAR	28,227,80	3,387	833.4160024
4	UTTAR DINAJPUR	3,000,849	3,140	955.6843949
5	DAKSHIN DINAJPUR	1,670,931	2,219	753.0108157
6	MALDA	3,997,970	3,733	1070.980445
7	MURSHIDABAD	7,102,430	5,324	1334.040195
8	NADIA	5,168,488	3,927	1316.141584
9	N.24 PARGANAS	10,082,852	4,094	2462.836346
10	S. 24 PARGANAS	8,153,176	9,960	818.5919679
11	HAORA	4,841,638	1,467	3300.366735
12	HUGLI	5,520,389	3,149	1753.060972
13	BARDDHAMAN	7,723,663	7,024	1099.610336
14	BIRBHUM	3,502,387	4,545	770.6022002
15	BANKURA	3,596,292	6,882	522.564952
16	PURULIA	2,927,965	6,259	467.8007669
17	PASCHIM MEDINIPUR	5,943,300	9,763	608.757554
18	PURBA MEDINIPUR	5,094,238	4,318	1179.767948
19	KOLKATA	4,486,679	185	24252.31892

Source: Provisional Census of India 2011

Fig. 2.2 Population Density of West Bengal (2011)

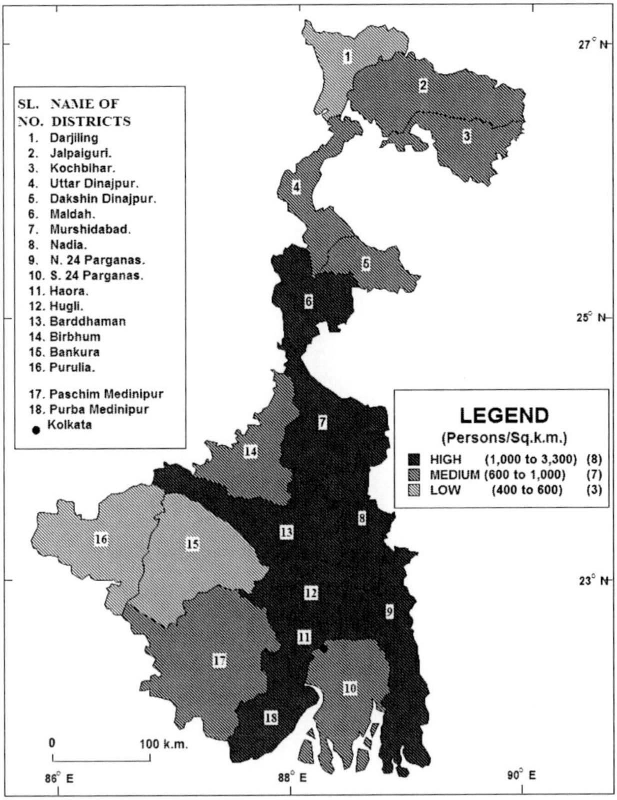

4. Analytical Methods

The present work uses the applied component method (Li et al. 2009), as used in assessing the ecological footprint of land use in China. This approach sums the ecological footprint of all relevant components of a population's resource consumption and waste production. In the present study we have divided the content of work into two parts: the ecological supply (or bio-productive areas) and the demand on nature (or ecological footprint).

Step 1: The computational forms can be defined as:

$$Ai=Ci/Yi,$$

where *i* is the item type of consumption, *Yi* is annual average yield of the item (kg/ha), *Ci*is per capita consumption of the item (kg/capita), and *Ai* is the per capita ecological footprint of the item(ha/capita).

Step 2:Calculate the ecological footprint of the studied region. The computational forms can be presented as:

ef= sum of (rj Ai)(j=1,2,…n),

where *ef* is per capita ecological footprint of the studied region (ha/capita), and *j* is the bio-productive area. The meaning of *i*, *Ai*, is similar to that in the first step, and *rj* is the equivalence factor (the equivalence factor of cropland is considered 2.9). Equivalence factor represents the world's average potential productivity of a given bio-productive area relative to the world average potential productivity of all bio-productive areas.

The total ecological footprint of the studied region can be defined as:

EF=N (ef),

where *EF* is the total ecological footprint (ha) and *N* is the population of the studied region.

Step 3: Calculate the biological capacity of the region. Bio-capacity is the counterpart of the ecological footprint, or the demand side. A region's total bio-capacity is the sum of its bio-productive areas. The computational form of per capita bio-capacity can be presented as:

ec=aj.rj.yj (j=1, 2,….n)

where *ec* is the per capita bio-capacity (ha/capita), *aj is* per capita bio-productive area, *rj* represents equivalence factor, and *yj* indicates yield factor. Yield factors describe the extent of a biologically productive area in a given country or region, which is more (or less) productive than the global average of the same bio-productive area. Each country or region has its own set of yield factors.

The total bio-capacity of the studied region can be defined as:

EC=N (ec),

where *EC* is the total bio-capacity of the studied region (ha) and *N* is the corresponding population.

Step 4: Calculate ecological deficit. A comparison of the footprint and bio-capacity reveals whether existing natural capital is sufficient to support consumption. An ecological deficit region indicates an area whose footprint exceeds its bio-capacity. The computational form of ecological deficit can be presented as:

Ecological deficit (ha) = Bio-capacity (ha) - Footprint (ha)

5. Results and Discussion

5.1 Yield Factor

The yield factor of different crops has been calculated through dividing the average yield of West Bengal by that of the world for the respective crop. Results show that the yield factors of pulse, sugarcane, jute, potato and tea are more than one, and cereals, oilseed and tobacco are less than one (see Table 2.2 below).

Table 2.2. Yield factor of different crops (2011)

Sl. No.	Name of Crops	Yield in West Bengal (Kg/Ha)	Average yield of World (Kg/Ha)	Yield factor
1	**Cereals**	2,728	3,380	0.807100592
2	**Pulse**	895	800	1.11875
3	**Total Oilseed**	1066	1,760	0.605681818
4	**Sugarcane**	72,786	59,500	1.223294118
5	**Jute**	2,732	1,300	2.101538462
6	**Potato**	35,768	17,200	2.079534884
7	**Tea**	2,255	1,072	2.103544776
8	**Tobacco**	1,477	1,856	0.795797414

Source: Directorate of Agriculture, Government of West Bengal, 2011, and F.A.O. calculated by the authors

5.2 Crop Production Characteristics

A wide variety of crops are produced in West Bengal. In the present study, eight major crops have been selected. Among these crops, cereals which include rice, wheat, millets etc. occupy a large area of 53,64,900 ha, but the productivity of these crops is lower in West Bengal than that of the world average. The area under pulse is 197, 100ha and the average productivity of pulse is higher than that of the world average. Oilseeds, which include mustered, til, mesta, sunflower etc., are produced in a 681,985 ha area but productivity is lower than the world average (see Table 2.3 below).

Table 2.3. Area of different crops in West Bengal (2011)

State/District	Area of cereals (000' Ha)	Area of pulses (000'Ha)	Area of total oilseed (Ha)	Area of sugarcane (Ha)	Area of jute(Ha)	Area of potato (Ha)	Area of tea (Ha)	Area of tobacco (Ha)
West Bengal	5,364.9	197.1	681,985	13,750	614,365	386,881	103,100	14,169
Darjeeling	62.1	1.1	497	3	2,980	7,600	32,501	--
Jalpaiguri	255.2	4.3	15,671	110	39,296	32,326	70,000	739
Koch bihar	295.6	5.6	12,220	-	87,208	27,877	199	13,180
Uttar Dinajpur	312.8	4.6	46,053	-	43,705	9,693	400	--
Dakshin Dinajpur	202.8	0.7	21,885	-	18,005	5,379	-	--
Malda	258.6	16.5	33,217	2,156	18,231	4,416	-	210
Murshidabad	404.5	50.8	99243	5266	171999	11899	-	-
Nadia	560.7	45.8	91733	1851	127607	5008	-	20
N.24 Parganas	231.2	7.8	50647	376	53654	7593	-	-
S. 24 Parganas	374.9	15.8	15503	76	2521	3008	-	-
Haora	115.8	1	8051	-	2996	6715	-	-
Hugli	293	0.4	49467	62	29040	91290	-	--
Barddhaman	564.3	3.2	46360	1070	12732	55166	-	-
Birbhum	282.2	17	38198	654	174	17671	-	-
Bankura	201.4	0.2	37202	10		32814	-	10
Purulia	160.7	9.8	2923	304		1187	-	10
Paschim Medinipur	622.4	5.6	91422	1617	3716	62679	-	-
Purba Medinipur	450	6.9	21693	195	501	4560	-	-
Kolkata	-	-	-	-	-	-	-	-

Source: Directorate of Agriculture, Government of West Bengal 2011

Sugarcane is produced in 13,750 ha and its productivity is higher than that of the world average. Jute is another important crop extensively produced in West Bengal. The districts in the lower Gangetic plain, like Nadia, Murshidabad, Barddhaman and Hugli, produce a large amount of jute. The average productivity of jute in West Bengal is twice the world average. Potato is produced in all the districts of West Bengal and the average productivity of this crop is twice that of the world average. Tea is another important crop produced in only four districts of West Bengal and here also the average productivity is twice the world average. In West Bengal, Darjeeling and Jalpaiguri are the leading tea producing districts. Tobacco is cultivated in five districts: Jalpaiguri, Kooch bihar, Malda, Nadia and Bankura (see Table 2.4 below).

Table 2.5 shows the consumption of major crops in different districts of West Bengal (except Kolkata). It clearly indicates a deficit in the supply of seven major crops like cereals, pulses, oilseeds, sugarcane, potato, tea, and tobacco and a surplus in jute supply in the state.

Table 2.4. Production of different crops in West Bengal (2011)

State/District	Production of cereals in 000' tonnes	Production of pulses in 000' tonnes	Production of total oilseeds in 000' tonnes	Production of sugarcane in 000' tonnes	Production of jute in tonnes	Production potato in 000' tonnes	Production of tea in 000' tonnes	Prod-uction of tobacco in tonnes
West Bengal	14,634.4	176.5	726.719	1,000.808	1,678,495.68	13,838.129	232.5	209,227
Darjeeling	136.3	0.7	0.242	0.296	6,628.5	149.374	75.2	-
Jalpaiguri	576.6	2.7	10.431	10.857	92,779.56	904.993	156.2	1,117
Koch bihar	730.5	4.1	6.328	-	195,141.06	845.019	0.4	19,644
Uttar Dinajpur	911.5	3.3	42.75	-	98,360.46	282.039	0.7	-
Dakshin Dinajpur	600.5	0.3	17.891	-	42,152.22	151.092	-	-
Malda	796.8	16.4	38.725	212.799	55,410.12	151.831	-	138
Murshidabad	1,165.4	45.7	128.141	335.123	466,126.56	370.671	-	-
Nadia	755.7	44.6	104.507	87.159	368,625.24	153.624	-	17
N.24 Parganas	626.7	7.1	66.381	17.705	197,073.36	254.965	-	-

S.24 Parganas	863.7	14.6	18.95	3.579	5557.5	63.657	-	-
Haora	305.4	0.7	10.667	-	982.98	242.614	-	-
Hugli	905.6	0.2	51.134	5.152	91,876.32	3,434.459	-	-
Barddhaman	1,669	3.2	41.761	86.488	39,454.38	2,268.242	-	-
Birbhum	836.4	17.1	37.738	64.978	549.9	613.243	-	-
Bankura	515.1	0.1	25.082	0.831	-	1,350.135	-	8
Purulia	276.6	2.6	1.92	25.261	-	24.924	-	3
Paschim Medinipur	1,732.3	4.1	93.85	134.367	7,677.72	2,448.136	-	-
Purba Medinipur	1,230.3	9.2	30.221	16.204	1,244.52	129.11	-	-
Kolkata	-	-	-	-	-	-	-	-

Source: Directorate of Agriculture, Government of West Bengal, 2011

Table 2.5. Consumption of major crops, West Bengal (2011)

Sl.no	State/ District	Consumption of cereals in 000' tonnes	Consumption of pulses in 000' tonnes	Consumption of oilseeds in 000' tonnes	Consumption of sugarcane in 000' tonnes	Consumption of jute in 000' tonnes	Consumption of potato in 000' tonnes	Consumption of tea in kg	Consumption of tobacco in kg
	West Bengal	16,637.61	511.12	2,649.084	2,125	102.0323	2,375.041	52,342,252.7	3,246,2418.5
1	**Darjiling**	335.49	10.3	53.418	42.857	2.4959	47.892	1,055,485.4	654,607.12
2	**Jalpaiguri**	704.8	21.65	112.22	90.015	4.3413	100.611	2,217,323.7	1,375,173.75
3	**Koch bihar**	514.12	15.79	81.86	65.667	3.1577	73.392	1,617,452.9	1,003,136.62
4	**Uttar Dinajpur**	546.55	16.75	87.024	69.805	3.3519	78.022	1,719,486.4	1,066,417.41
5	**Dakshin Dinajpur**	304.33	9.34	48.456	38.874	1.8625	43.444	957,443.4	593,801.99
6	**Maldah**	728.17	22.37	115.941	93.061	4.4634	103.947	2,290,836.8	1,420,766.13
7	**Murshidabad**	1293.6	39.75	205.97	165.227	7.9983	184.663	4,069,692.3	2,524,004.07
8	**Nadia**	941.36	28.96	149.886	120.293	5.7619	134.38	2,961,543.6	1,836,735.72
9	**N.24 Parganas**	1,836.44	56.41	292.402	234.556	11.5124	262.154	5,777,474.1	3,583,162.88
10	**S. 24 Parganas**	1,484.97	45.62	236.442	189.668	9.6116	211.982	4,671,769.8	2,897,409.44
11	**Haora**	881.83	27.09	140.407	112.623	5.3967	125.882	2,774,258.5	1,720,582.22
12	**Hugli**	1,005.45	30.85	160.091	128.418	6.1289	143.53	3,163,182.8	1,961,791.91
13	**Barddhaman**	1,406.75	43.25	223.986	179.679	8.6262	200.815	4,425,658.8	2,744,772.97
14	**Birbhum**	637.9	19.55	101.569	81.435	3.9115	91.062	2,006,867.7	1,244,649.53
15	**Bankura**	655.01	20.14	104.292	83.694	4.0169	93.503	2,060,675.3	1,278,021.48
16	**Purulia**	533.28	16.25	84.91	68.136	3.2704	76.127	1,677,723.9	1,040,516.35
17	**Paschim.Medinipur**	1,082.48	33.25	172.355	138.228	6.6387	154.525	3,405,510.9	2112081.87
18	**PurbaMedinipur**	927.83	28.51	147.732	118.571	5.6853	132.45	2,918,998.3	1,810,349.22
19	**Kolkata**	817.18	25.70	130.11	104.37	4.89	116.65	2,570,866.988	1,594,440.277

Source: Food and Supply Department, Government of West Bengal 2011

6. Characteristics of Cropland Footprint and Cropland Bio-capacity

Ecological footprint and bio-capacity of cropland of different districts of West Bengal have been measured here (see Table 2.6 and Fig. 2.3 below).

Table 2.6. Ecological surplus and/or deficit of croplands of West Bengal (2011)

State/District	Total ecological footprint of cropland (ha)	Total bio-capacity of cropland (ha)	Ecological surplus(+) or deficit (-) of cropland (ha)
WEST BENGAL	27,066,449.56	21,182,183.28	-5,884,266.279
DARJEELING	821,972.2718	412,062.8096	-409,909.4622
JALPAIGURI	1,517,177.955	1,502,345.658	-14,832.29681
KOCH BIHAR	1,142,192.317	1,462,744.753	320,552.436
UTTAR DINAJPUR	903,581.5031	1,155,205.917	251,624.414
DAKSHIN DINAJPUR	1,585,456.908	2,316,330.57	730,873.6625
MALDA	1,063,955.695	863,026.7597	-200,928.935
MURSHIDABAD	1,936,278.025	2,424,583.096	488,305.0707
NADIA	2,529,245.572	2,436,599.212	-92,646.3601
N.24 PARGANAS	2,851,726.335	1,029,528.088	-1,822,198.247
S. 24 PARGANAS	2,637,966.313	989,753.577	-1,648,212.736
HAORA	1,415,527.428	347,181.1237	-1,068,346.304
HUGLI	1,600,664.592	1,501,720.04	-98,944.55185
BARDDHAMAN	2,265,357.38	1,826,685.66	-438,671.7202
BIRBHUM	997,393.9157	892,711.449	-104,682.4668
BANKURA	1,329,206.669	735,336.8552	-593,869.8135
PURULIA	1,480,963.562	421,322.0641	-1,059,641.498
PASCHIM MEDINIPUR	1,780,443.858	2,041,912.046	261,468.1881
PURBA MEDINIPUR	1,385,380.706	1,145,000.589	-240,380.1173
KOLKATA	1,329,408.655	0	-1,329,408.655

Source: Calculated by the authors

The results show that the footprint zones have resemblance to the number of population in the corresponding districts. Highly populated districts have higher cropland footprints while districts with lower populations have a low footprint of cropland (see Fig. 2.3 below).

Fig. 2.3: Crop Footprint Pattern of West Bengal (2011)

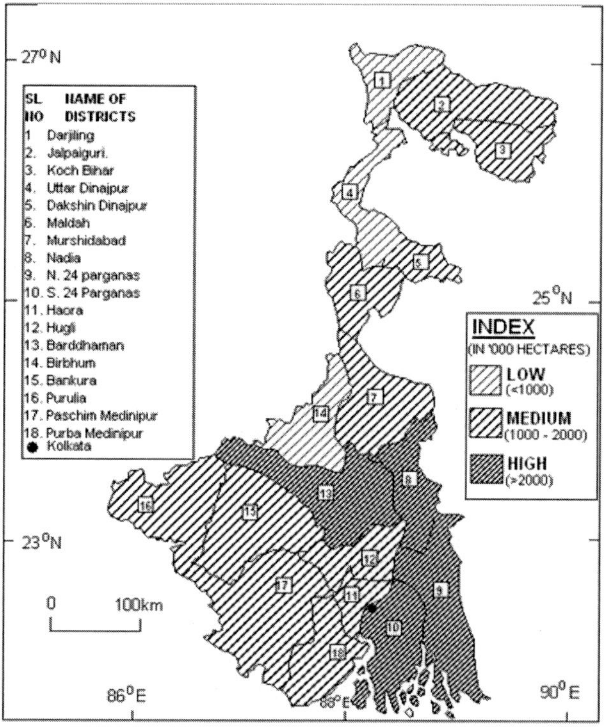

Cropland bio-capacity is very much related to the amount of cropland and yield factor. The higher the area of cropland and yield per hectare, the higher the cropland bio-capacity. The districts of Darjeeling, Malda, Purulia, Bankura, Haora and South 24 Parganas have less cultivable land and low cropland bio-capacity (less than 1,000,000 ha). On the other hand, the districts of Dakshin Dinajpur, Murshidabad, Nadia and Paschim Medinipur, with a large amount of cultivable land, have high cropland bio-capacities (more than 2,000,000 ha). Kolkata has hardly any bio-capacity at all (see Fig. 2.4 below).

Fig. 2.4: Pattern of Cropland Biocapacity in West Bengal (2011)

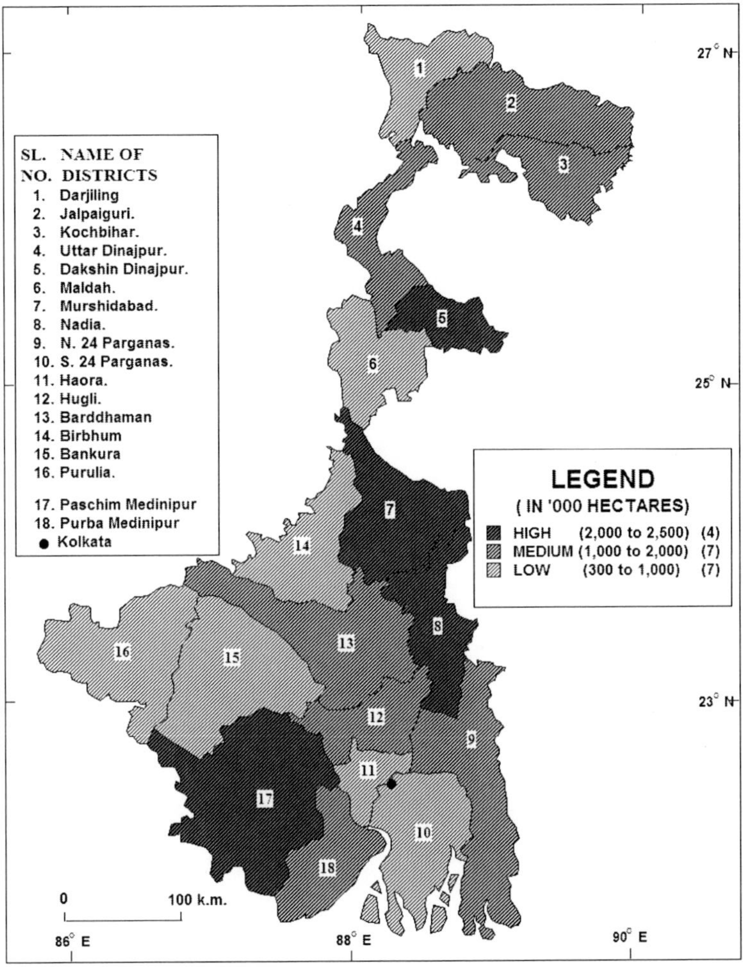

The districts with more bio-capacity of cropland than footprint are experiencing ecological surplus, and the districts with more footprint of cropland than bio-capacity are experiencing ecological deficit. In West Bengal, most of the districts are suffering from the ecological deficit of cropland. In the districts of Darjeeling, Jalpaiguri, Malda, Birbhum, Nadia, Barddhaman, Hugli and Purba Medinipur, the ecological deficit of cropland is low (less than 500,000 ha) and in the districts of North and South 24

Parganas, Kolkata, Haora, Purulia and Bankura, it is high (more than 500,000 ha). Therefore, the croplands of these districts are unable to supply the required food for the population. On the other hand, the districts of Kooch Bihar, Uttar Dinajpur, Murshidabad and Paschim Medinipur are experiencing low ecological surplus of cropland (less than 500,000 ha),while ecological surplus in the district of Dakshin Dinajpur is very high (more than 500,000 ha). The low population density and a large extent of alluvial tract have created a very good agricultural environment here (see Fig. 2.5 below).

Fig. 2.5 Ecological Surplus or Deficit of Cropland in West Bengal (2011)

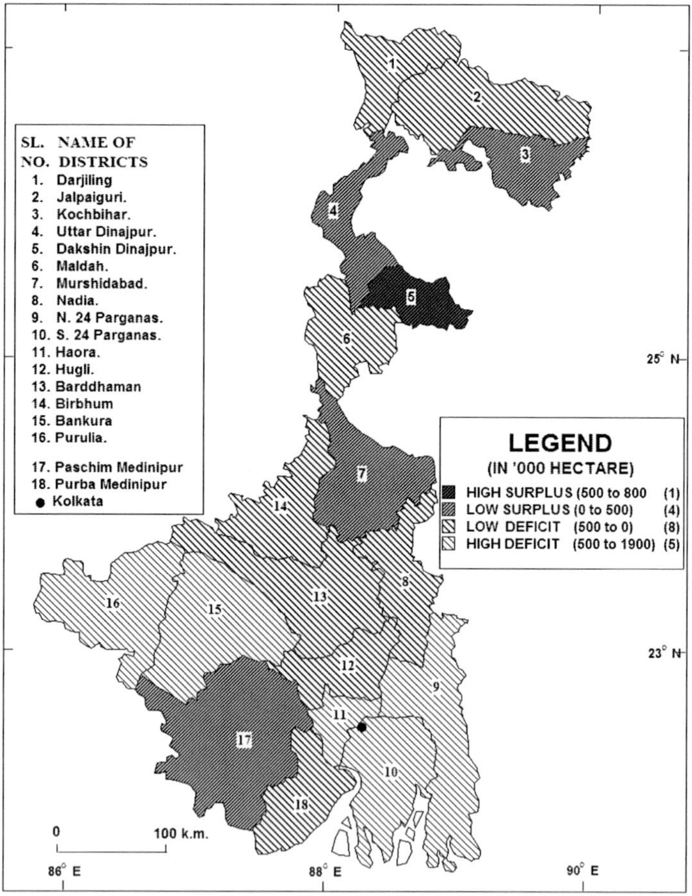

7. Conclusion

It can be stated that we are living beyond our biophysical means (Moffatt 2000). Except for the five districts of Kooch Bihar, Uttar Dinajpur, Dakshin Dinajpur, Murshidabad and Paschim Medinipur, the remaining fourteen are experiencing ecological deficits of croplands (ha). The situation is crucial in the nine districts of Malda, Murshidabad, Nadia, North 24 Parganas, Haora, Hugli, Barddhaman, Purba Medinipur and Kolkata with a high density of population.

The main hindrance lies in the adoption of such an agricultural system which is neither economically sound nor environmentally sustainable. Application of chemical fertilizer instead of biofertilizer, and the use of fertilizer without examination of soil nutrient, is deteriorating the soil health day by day. At the same time yield per hectare of crops needs to increase to manage the growing demand of food for the geometrically increasing population figure of the state. Therefore, we are now at the threshold of thinking about restricting population size, the rational use of land and the implementation of a proper agricultural system for environmental and economic sustenance of the state.

References

Ayres, R. U. "Commentary on the Utility of the Ecological Footprint Concept." *Ecological Economics* 32 (2000): 347–349.

Basu, A. K. "Agriculture of West Bengal." In *West Bengal*, eds. A. B. Chatterjee, A. G. Gupta and P. K. Mukhopadhyay, 85–95. Calcutta: Firma K.L Mukhopadhyay, 1970.

Bose, S. C. *India—The Land and The People series—Geography of West Bengal.* New Delhi, India: National Book Trust, 1979.

Costanza, R. "The dynamics of the Ecological Footprint Concept." *Ecological Economics* 32 (2000): 341–345.

Hanley, N., I. Moffatt, R. Faichey & M. Wilson. "Measuring Sustainability: A Time Series of Alternative Indicators for Scotland." *Ecological Economics* (1999): 55–73.

Herberl, H., K. H. Erb & F. Krausmann. "How to Calculate and Interpret Ecological Footprints for Long Periods of Time: The Case of Austria 1926–1995." *Ecological Economics* 38 (2001): 24–45.

Kathryn. "New Methodology for the Ecological Footprint with an Application to the New Zealand Economy." *Ecological Economics* 27 (1998): 149–160.

Lenzen, M., & S. A. Murray. "The Ecological Footprint: Issues and Trends." *The University of Sydney: Integrated Sustainability Analysis, NSW*, n.d.

Levett, R. "Foot Printing: A Great Step Forward but Tread Carefully." *Local Environment* 3 (1) (1998): 67–74.

Li, M. & B. Li. "Assessment of Ecological Footprint of Land Use in Chongqing, China." In *Remote Sensing and Geoscience for Agricultural Engineering*. 2009.

Moffatt, I. "Ecological Footprints and Sustainable Development." *Commentary Forum: The Ecological Footprint, Ecological Economics* 32 (2000): 359–362.

Muniz, I., & A. Galindo. "Urban Form and the Ecological Footprint of Commuting: The Case of Barcelona." *Ecological Economics* 55 (2005): 499–514.

Rapport, D. J. "Ecological Footprints and Ecosystem Health: Complementary Approaches to a Sustainable Future." *Ecological Economics* 32 (2000): 381–383.

Rees, W. E. "Ecological Footprints and Appropriated Carrying Capacity: What Urban Economics Leaves Out." *Environment and Urbanisation* 4 (2) (1992): 121–130.

Stoglehner, G. "Ecological Footprint-a Tool for Assessing Sustainable Energy Supplies." *Journal of Cleaner Production* (2001): 267–277.

Wackernagel, M. & W. E. Rees. *Our Ecological Footprint—Reducing Human Impact on the Earth.* New Society Publishers Philadelphia, PA, USA, 1995.

Wackernagel, M. et al. "National natural capital accounting with the ecological footprint concept." *Ecological Economics* 29 (3) (1999).

CHAPTER THREE

COST AND RETURNS OF MAJOR CROPPING SYSTEMS: A CASE STUDY IN THE DISTRICT OF BURDWAN IN WEST BENGAL

NIRMALENDU SARKAR, SANTOSH KUMAR DUTTA AND SWAPAN KUMAR BISWAS

1. Introduction

In agriculture, just as in all branches of material production, the essence of economic activity is production. Accordingly, in connection with economic activity in agriculture, the cropping system is an important determinant. Cropping pattern means the proportion of area under different crops at a point of time, while cropping system refers to the crops and crop sequences and the management techniques used on a particular field over a period of years. In a bigger system, every subsystem is not only interlinked but also interdependent. Farmers are part of the system and are able to set or modify their own goals. Therefore, two farms with identical climate and soil may be managed with different aims to achieve a different output. In fact, the ultimate profit of any agro-climate zone depends on the type of cropping system adopted. A variety of cropping systems are well suited to Burdwan due to its diverse climatic conditions, availability of irrigation facilities and soil types. The cropping systems in any area are influenced by physical, social and economic factors. The other determinants of cropping systems include land type, market and related prices of farm products, labour and capital resources. Farmers manage their available resources to obtain maximum profit by adopting a suitable farming/cropping system. Hence, the present study has been undertaken with the main objective of analyzing costs and returns in major cropping systems of four selected villages, and also to assess the remunerative cropping systems in the Burdwan district of West Bengal.

2. Materials and Method

2.1 Geographical location

The Burdwan district is known as a rice growing area—"the store-house of paddy-crop of the state." The district is situated between latitudes 22°56' to 23°53' North and longitudes 88°48' to 88°25' East. Lying within the Burdwan Division, the district is bounded on the north by Dumka of Jharkhand, Birbhum and Murshidabad; on the east by Nadia; on the south by Hooghly, Bankura and Purulia; and on the west by Dhanbad (of Jharkhand) districts.

2.2 Data Source

No secondary data have been generated in this study, and the whole study is based on primary data. The field level information has been collected from the respondent farmers during the agricultural year 2005–2006. There are six sub-divisions in the Burdwan district. Among the sub-divisions, the people of two sub-divisions Asansol and Durgapur are highly engaged in industrial activities. Our study is mainly concerned with agricultural activities. For this reason these two subdivisions have not been considered for the study. As these two sub-divisions have been left out, the study is mainly concerned with the remaining four sub-divisions. Survey work of the study has been carried out with primary data pertaining to four villages selected from the remaining four sub-divisions, taking one each from one sub-division. Considering the easy accessibility and familiarity with the farmers, these four villages have been selected purposively. We have taken different categories of farmers randomly from the sample villages.

The administrative set-up, economy and communication of the selected villages are shown in the Table 3.1(a). The village Nashigram is under the Sadar (N) sub-division. The economy of the village is good and well connected to the headquarters and other commercial places. The village Kashiyara under the Sadar (S) sub-division is a village whose economy is good but communication is bad. The village Hapania under the Kalna sub-division has a bad economic condition but good communication. The village Chhoto-Maliha under the Katwa sub-division is a village with a bad economy and bad communication. We have taken an equal number of respondent farmers from all sample villages for the study, i.e. out of a total of two hundred farmer-families [table 3.1(b)], 50 farmers are from the village Nashigram of the Sadar (North) sub-division, another 50 are from the village Kashiyara of the Sadar (South) sub-division, 50 are from the village Hapania located within the Kalna sub-division, and 50 are from the village Chhoto-Maliha of the Katwa sub-division. We have taken different categories of farmers randomly from the sample villages. Table-1(b) clearly shows the number of marginal, small, semi-medium, medium and large farmers of the four selected villages separately. Farmer families have been classified into five categories on the basis of land holdings as follows:

(1) Marginal: Who have land varying in size below 1 acre
(2) Small: Who possesses lands varying in size from 1 acre to below 2 acres
(3) Semi-Medium: Who own land between 2 acres to below 5 acres
(4) Medium: Who have lands between 5 acres and below 10 acres
(5) Large: Who have land of 10 acres and above.

The survey aims to collect all relevant information relating to farming activities of respondent farmers, e.g. size of holdings, quality and quantity of seeds, total expenditure on seeds, area under HYV, cost of various types of fertilizers used, sources of irrigation, expenditure in irrigation, nature of machinery used and expenditure related to their use, loans taken from formal and informal sources, interest on loans, loan advances to workers, storage costs, crop-hoarding periods, crop-shrinkage, marketing facilities, transportation costs, types of crops cultivated by the respondent farmers and proportion of area cultivated under each crop by the respondent farmers. We have also collected sale price and quantity sold of different crops separately from each respondent farmer.

2.3 Methodology

A simple tabular analysis technique was used to estimate the costs and returns in different cropping systems of the sample villages in the Burdwan district.

3. Results and discussion

A comparative advantage of agriculture produce is usually expressed in terms of net return per acre of land. By total comparative advantage of agriculture of any cropping system, we mean the income from sale proceeds of a specific cropping system, plus the imputed value of the crop or crops of the specific cropping system retained by the family concerned, minus total expenditure on that cropping system produced in an agricultural year. After that, we divide the total comparative advantages by total land size of the specific cropping system and get the net return or net profit per acre of land. The main objective of a farm is to get the maximum net return per acre of land in an agricultural year by making best utilization of its land. The net return or net profit per acre of land of a farm depends on the type, method and intensity of production, as well as the economic aspirations of the farmer and the social context within which the farming is carried out. The cultivation of different possible combination of crops, grown in successive seasons on the same plot or plots, plus other crops on other plots in an agricultural year, are taken into consideration in the case of calculating the total production of a farm and also the total income of the cropping systems at a point of time.

Both production costs and annual net returns must be considered in choosing suitable cropping systems. To examine the remunerative cropping systems we begin by analyzing (a) the area under HYV crops and expenses on HYV crops, expenses on chemical & bio fertilizer & pesticides, and levels mechanization per acre, and (b) concluding with the details of all cropping practices including their costs and returns & BCR (Benefit Cost Ratio) in all the sample villages.

(a) (i) Area under HYV crops and Expenses on HYV crops:

Tables 3.2(a), 3.2(b), 3.2(c) and 3.2(d) clearly show the percentage of area under HYV crops and expenses on HYV crops per acre of land for different size categories of farmers in the different sample villages. We observe from the tables that the percentages of area under HYV crops of the respondent marginal farmers in Kashiyara (Table 3.2[b]) are highest (57.41%) and lowest (43.94%) in Chhoto-Maliha (Table 3.2[d]). Among the different size categories of respondent farmers in Nashigram (Table 3.2[a]) the percentages of area under HYV crops are higher (90.12%) for large groups and lower (54.72%) for small farmers. Similarly, expenses per acre of land are the highest (Rs. 1,903.88 for all groups taken together) in Kashiyara (Table 3.2[b]) and the lowest (Rs. 332.26) in Chhoto-Maliha (Table 3.2[d]). The other information relating to the different groups of the same village and same groups of the different villages is clearly shown in Tables 3.2(a), 2(b), 2(c) and 2(d) also. HYV crops are more intensively cultivated in two advanced villages, namely Kashiyara and Nashigram. Cultivation of HYV crops is more expensive and hence the farmers, who are financially better off, usually cultivate HYV crops in relatively greater percentages of the gross cropped area.

(a) (ii) Fertilizer intensity:

Production of a crop depends on fertility of soil, and fertility depends on fertilizer intensity, i.e. the use of chemical fertilizer as well as bio-fertilizer per unit of land. Use of chemical and bio-fertilizer mainly depends on the availability of the irrigation and intensity of HYV crops. For cultivating HYV crops more fertilizers are needed per unit of land compared to traditional crops. From Tables 3.3(a), 3.3(b), 3.3(c) and 3.3(d) it may be observed that total fertilizer expenses per acre are the highest (Rs. 1,648.82) in Kashiyara and the lowest (Rs. 619.15) in Hapania. Bio-fertilizer is not used in Hapania. It is noted that the scarcity of cow dung, due to its use mainly as fuel for cooking, is the reason for the low use of bio-fertilizer in that village. Organic matters, such as residues of crop plants, are hardly used as this is mainly fed to cattle. Again, lack of adequate knowledge about bio-fertilizer is the root cause for the low use or absence of bio-fertilizer use by the respondent farmers in the sample villages.

Pesticides are generally used to destroy the insects which are harmful to plants. For the vertical expansion of land farmers generally adopt HYV crops, but these are very much susceptible to pests and insects and the application of pesticides and insecticides becomes essential. Expenses on pesticides per acre of land are the highest (Rs. 162.62) in Nashigram and the lowest (Rs. 17.53) in Hapania. In Nashigram, pesticide expenses per acre are higher (Rs. 214.60) for medium groups and lower (Rs. 144.57) for small farmers. Tables 3.3(a), 3.3(b), 3.3(c) and 3.3(d) also depict expenses on total fertilizer, bio-fertilizer, chemical fertilizer and pesticides for different size categories in the same village, and also the same groups for the different sample villages. It is also notable from our sample survey that small size farms usually spend very small amounts on chemical fertilizer and pesticides, and as a result their agriculture practice is more eco-friendly and sustainable.

(a) (iii) Mechanization:

By mechanization of agriculture we mean the replacement of animal and human power by machinery. The purpose of machinery is to raise agricultural productivity, increase profitability in agriculture and improve quality of life of the farming population. As regard to the levels of mechanization in the four sample villages, Table 3.4 clearly shows the number of major tools and machinery used per acre of land in the sample villages. The major agricultural tools are Tractor, Pumpset, Submersible, Sprayer, Duster and Thresher. In Table 3.4 we observe that the number of tractors used per acre is highest (0.0533) in Kashiyara and lowest (0.0109) in Chhoto-Maliha. We find that expenses on modern implements per acre are the highest (Rs. 1182.58) in Kashiyara and the lowest (Rs. 353.19) in Chhoto-Maliha. Again, the expenses on submersible per acre are the highest (Rs. 2500) in Chhoto-Maliha and the lowest (Rs. 1500) per acre in Kashiyara. We found that in our sample villages the extents of irrigated land as a percentage of total land are 98.06, 97.21, 100 and 52.9 respectively. Again, Chhoto Maliha is still beyond the purview of electrification. For this reason, submersibles are functioning with the help of diesel and as a result the cost of water pumping is highest in Chhoto-Maliha among the sample villages.

The empirical results, as presented in Tables 3.5 to 3.8, reveal the net return per acre of land from different cropping systems adopted by the respondent farmers of the sample villages Nashigram, Kashiyara, Hapania and Chhoto-Maliha, respectively, in the agricultural year 2005–06. The results in Table 3.5 clearly show that the respondent farmers of Nashigram followed 12 cropping systems in the agricultural year 2005–06, and we have described those cropping systems as group I, group II, group III etc. It further reveals that 42% of the total respondent farmers of Nashigram adopted the group VI cropping system. The second highest group of 18% were engaged in the cultivation of the group II cropping system. It is also noted that the respondent farmers of the group VI cropping system use the highest percentage (34.27%) of the net sown area for multi-crop cultivation. However, MCI of 1.942 is the highest for the cropping system group V. It is further observed that the average annual return per acre of Rs. 23,939.55 is the highest for the cropping system group V among all the groups. The average yearly production costs of the cropping system group X is Rs. 5680.26, which is the lowest among all the cropping systems. It is also noted from the table that the average yearly net profit of Rs. 11, 591.96 per acre was earned by the farmers of cropping system group VI and is incidentally the highest net profit per acre among all the cropping systems groups.

The details of per acre cost and returns of the village Kashiyara are given in Table 3.6. It is observed that the respondent farmers of the sample village Kashiyara have adopted seven cropping systems during the agricultural year 2005–06. It is also revealed from the table that 26% of the respondent farmers, who constitute the largest group, are engaged in the group II cropping system. Another 26% of farmers also followed the group IV cropping system. The second highest percentage (24%) is noted in the case of the group V cropping system. It is also noted that the respondent farmers of the group V cropping system use the highest percentage (38.67%) of the net sown area. The highest MCI (2.857) is found for the cropping system group IV. From the table it transpires that average annual return per acre is the highest (Rs. 54,569.35) for the cropping system group IV, and the average yearly production cost is the lowest, i.e. Rs. 14,960.54, which is noted in the case of the cropping system group III. It is also observed that average yearly net profit per acre is the highest (Rs. 22, 669.34) for the cropping system group VII.

Table 3.7 presents different cropping systems which have usually been followed by the respondent farmers of the sample village Hapania. It may be noted from the table that 21 different cropping systems were generally adopted and prescribed by the farmers of Hapania during the agricultural year 2005–06 and we expressed those types of cropping systems as group I, group II, group III etc. However, 26% (the highest percentage) of the respondent farmers of Hapania are engaged in the group III cropping system, and the second highest at 20% adopted the group II cropping system. It is also noted that the respondent farmers of the group III cropping system used the highest percentage (36.09%) of the net sown area. The highest MCI is 3.0, attained by the cropping system groups X, XIV and XX. From Table 3.7 it may further be noted that the highest average annual return per acre is Rs. 37,850 earned by the cropping system group XX, and the average yearly production cost is the lowest at Rs. 11,016.67, attained by the farmers who adopted the cropping system group XXI. It is further revealed that the highest average yearly net profit per acre is Rs. 17,393.33, earned by the farmers of cropping system group XVI.

Table 3.8 shows that the respondent farmers of Chhoto-Maliha normally followed 17 cropping systems during the agricultural year 2005–06, and we described these categories as group I, group II, group III, etc. It is noted from the table that the majority of the respondent farmers (20%) are engaged in the group V cropping system, and 16% of the respondent farmers prefer to adopt the group VIII cropping system. It is also noted that the respondent farmers of the group V cropping system use the highest percentage (19.23%) of the net sown area. The highest MCI is 2.0, attained by the cropping system groups VII and X. From the table it transpires that average annual return per acre is the highest (Rs. 21288.33), earned by adopting the cropping system group III, and the average yearly production cost is the lowest Rs. 3691.67) in case of the cropping system group VII. It is also observed that the highest average yearly net profit per acre of land of Rs. 12181.88 earned by adopting the cropping system group III.

Table 3.9 shows an overall agricultural performance of the respondent farmers in the different villages of the Burdwan district of West Bengal in the agricultural year 2005–06. The result shows that the respondent farmers of Kashiyara have the highest Physical Productivity of 4,157 kg per acre and the lowest of 1,642 kg per acre, produced by the respondent farmers of Chhoto-Maliha. MCI is thehighest (2.365) for Kashiyara and the lowest (1.451) in the case of Chhoto-Maliha. Again, Kashiyara also enjoys the highest per acre effective value productivity of Rs. 18,942, and Chhoto-Maliha secures the lowest effective value productivity of Rs. 9,830 per acre. It is also revealed from Table 3.9 that the highest production cost of Rs. 28,067 per acre of land cultivated is recorded for Kashiyara, but Chhoto-Maliha has achieved the lowest production cost of Rs. 6,274 per acre. Again, Kashiyara has the highest average annual net profit of Rs. 16734 per acre. Though the value productivity (Rs. 18942) per acre of Kashiyara is greater than that of the remaining three villages, we observed from Table 3.9 that Chhoto-Maliha recorded the highest (1.279) Benefit Cost Ratio (BCR). This is due to the fact that Chhoto-Maliha enjoys its lowest production cost on the one hand (particularly for expenses on HYV seed and modern implements, and plant protection measures, per acre), and on the other returns earned by the farmers in this village are the lowest because of the low average annual value productivity per acre, by adopting various types of cropping systems.

4. Conclusion

The questions of sustainable income and the employment of farmers have been important issues in the days of globalization, calling for the remunerative cropping systems. To examine the remunerative cropping systems, this chapter explained the costs and returns of the major cropping systems for the agricultural year 2005–06 in the sample villages chosen from the district of Burdwan of West Bengal. From the preceding analysis and observations it may be noted that not only the physical factors of production but also numerous socio-economic factors like price of agricultural produce, production cost per acre, MCI, irrigation facilities, communication system, storage facilities, choice of cropping system, economic conditions and land size have a direct or indirect bearing on the average net return per acre of land in an agricultural year.

Net returns were directly related to the price that the producer received for the product and inversely related to the production costs. Annual production costs increased with increasing cropping intensity. Net return was the highest for potato, as well as potato containing cropping systems, mainly due to the higher yield of potato in Kashiyara. Though growing potato was associated with high production costs it was also the most profitable crop. The results revealed that the cropping system VII (aman, boro, potato, mustard and pumpkin) was considered the most profitable among all systems studied in Kashiyara, and also among the other three villages, as it contributed the highest returns per acre.

In Hapania and Chhoto-Maliha it was found that the respondent farmers are interested in producing traditional crops because the traditional crops require fewer input costs and are also less risky, which makes them more suitable.

BCR can be used to evaluate the economic merit of a farm. Ultimately, BCR aims to examine the potential actions of a farm with the objective of increasing net returns. But, when BCR of a cropping system category is the highest we cannot come to the conclusion that the respondent farmers of that cropping system category enjoy the strongest force in promoting net return per acre in an agricultural year. We observed such a situation in Chhoto-Maliha, where agricultural performance is very poor. Chhoto-Maliha recorded the highest BCR due to its lowest production cost, particularly for seed, expenses on modern implements and plant protection measures per acre. It was also noted that the net average returns earned by the farmers in this village by adopting various types of cropping systems were also the lowest because of low average annual value productivity per acre.

The poor economic condition of the villages Hapania and Chhoto-Maliha was the reason for not using machines. Secondly, in all the sample villages it was found that the small size of land holding was the other constraint inhibiting the use of machines. It is also noted that electricity used for agricultural purposes, marketing infrastructure and transport facilities are crucial for increasing agricultural value from the village Chhoto-Maliha. Thus, the operation and interaction of a variety of factors are usually observed to influence the productivity and profitability in the agricultural scenario.

References

Biswas, B., D. C. Ghosh, M. K. Dasgupta, N. Trivedi, J. Timsina, and A. Dobermann. "Integrated assessment of cropping systems in eastern Indo-Gangetic plain." *Field Crops Research* 99 (2006): 35-47.

Chakarabarti, B. "Economic Development of the district Burdwan and Bankura—A Comparative study." *Published Ph. D. Thesis*, 1995.

De, U. K. "Crop Diversification in West Bengal during 1970-71 to 1994-95." *Published Ph. D. Thesis*, 2000.

De, U., and M. Chattopadhyay. "Crop diversification by poor peasants and role of infrastructure: Evidence from West Bengal." *Journal of Development and Agricultural Economics* 2, no. 10 (2010): 340-350.

Dhanaji, K. Lahu, and K. Nita Dhanaji. "Crop Concentration in Sindhudurg District -A Geographical Analysis." *Geo Science Research* 1, no. 2 (2010): 28-33.

Dyer, G. *Class, State and Agricultural Productivity in Egypt.* Midsomer North Avon: Great Britain Book craft Ltd., 1995.

Ghosh, R. K. "Environment and Pesticides." *B.C.K.B., Mohanpur*, 2006.

Husian, M. *Agricultural Geography.* Rawat Publications, 2002.

Kumara, B. R., S. B. Hosamani, N. R. Mamaledesai, S. N. Megeri, and M. H. Hosamni. "Costs and Returns of major cropping systems in northern transition zone of Karnataka." n.d.

Leong, G. C., and G. Morgan. *Human and Economic Geography.* Oxford university Press, 2004.

Table 3.1(a). Administrative set up, socio economic condition and communication of the sample villages

Descriptions	Nashigram	Kashiyara	Hapania	Chhoto-Maliha
District	Burdwan	Burdwan	Burdwan	Burdwan
Sub-Division	Sadar North	Sadar South	Kalna	Katwa
Block	Bhatar	Memari-I	Purbasthali-II	Ketugram - II
Panchayet	Barbaloon-II	Gope-gantar-II	Pila	Billeswar
Economy	Good	Good	Bad	Bad
Communication	Good	Bad	Good	Bad

Source: Field Survey

Table 3.1(b). No. of marginal, small, semi-medium, medium and large farmers of the sample villages

Descriptions	Marginal	Small	Semi-medium	Medium	Large	Total
Nashigram	9	7	9	11	14	50
Kashiyara	3	12	10	20	5	50
Hapania	12	12	19	7	0	50
Chhoto-Maliha	7	12	17	12	2	50
Total	31	43	55	50	21	200

Source: Field Survey

Table 3.2(a). Area under HYV crops according to different size group of respondent farmers in Nashigram

Descriptions	Area under all crops (in acres)	Area under HYV crops (in acres)	Exp. on HYV crops per acre (in Rs.)
Marginal	11.22	6.32 (56.32)	328.43
Small	15.46	8.46 (54.72)	227.03
Semi - Medium	37.06	20.66 (55.74)	165.67
Medium	97.2	76.0 (78.19)	310.03
Large	445.2	401.2 (90.12)	275.92
Total	606.14	512.64 (84.57)	279.35

Table 3.2(b). Area under HYV crops according to different size group of respondent farmers in Kashiyara

Descriptions	Area under all crops (in acres	Area under HYV crops (in acres)	Exp. on HYV crops per acre (in Rs.)
Marginal	5.4	3.1 (57.41)	2162.03
Small	38.74	28.4 (73.31)	1355.55
Semi-Medium	70.6	47.96 (67.93)	1648.51
Medium	273.24	229.64 (83.92)	1842.19
Large	156.4	136 (86.96)	1678.58
Total	544.38	446.38 (81.76)	1738.58

Source: Field Survey. Note: Figures in the parentheses are in Percentages

Table 3.2(c). Area under HYV crops according to different size group of respondent farmers in Hapania

Descriptions	Area under all crops (in acres)	Area under HYV crops (in acres)	Exp. on HYV crops per acre (in Rs.)
Marginal	14.46	6.6 (46.05)	717.22
Small	37.4	18 (49.73)	516.36
Semi-Medium	90.4	58 (64.93)	465.87
Medium	71	56 (79.01)	351.69
Large	---	---	---
Total	213.26	140.0 (65.68)	461.78

Table3.2(d). Area under HYV crops according to different size group of respondent farmers in Chhoto-Maliha

Descriptions	Area under all crops (in acres)	Area under HYV crops (in acres)	Exp. on HYV crops per acre (in Rs.)
Marginal	6.6	2.9 (43.94)	143.64
Small	23.1	6.2 (26.84)	73.68
Semi-Medium	69.64	18.1 (25.99)	62.12
Medium	101.98	23.9 (23.44)	67.33
Large	62.8	22.4 (35.67)	74.68
Total	264.12	73.5 (27.83)	70.16

Source: Field Survey.
Note: "---" Means does not exist; Figures in the brackets are in Percentages.

Table 3.3(a). Class-wise expenses per acre of land on chemical fertilizer, bio-fertilizer, total fertilizer and pesticides of respondent farmers in Nashigram

Descriptions	Chemical Fertilizers (Rs.)	Bio Fertilizers (Rs.)	Total Fertilizers (Rs.)	Pesticides (Rs.)
Marginal	707.48	49.02	756.51	196.97
Small	521.28	148.77	670.05	144.57
Semi-Medium	757.28	180.51	937.80	192.76
Medium	735.87	155.04	890.92	214.60
Large	762.39	128.32	890.71	146.90
All Gr.	750.95	134.85	885.82	162.62

Table 3.3(b): Class-wise expenses per acre of land on chemical fertilizer, bio-fertilizer, total fertilizer and pesticides of respondent farmers in Kashiyara

Descriptions	Chemical Fertilizers (Rs.)	Bio Fertilizers (Rs.)	Total Fertilizers (Rs.)	Pesticides (Rs.)
Marginal	1445.00	274.26	1719.26	129.07
Small	1398.71	138.22	1536.93	146.64
Semi-Medium	1350.75	460.97	1525.68	123.57
Medium	1517.56	168.83	1689.39	102.86
Large	1559.97	124.36	1684.72	93.28
All Gr.	1492.00	156.82	1648.82	106.81

Table 3.3(c). Class-wise expenses per acre of land on chemical fertilizer, bio-fertilizer, total fertilizer and pesticides of respondent farmers in Hapania

Descriptions	Chemical Fertilizers (Rs.)	Bio Fertilizers (Rs.)	Total Fertilizers (Rs.)	Pesticides (Rs.)
Marginal	603.94	0	603.94	33.47
Small	598.88	0	598.88	37.03
Semi-Medium	666.57	0	666.57	32.63
Medium	572.53	0	572.53	25.63
Large	----	----	----	--
All Gr.	619.15	0	619.15	17.53

Table 3.3(d). Class-wise expenses per acre of land on chemical fertilizer, bio-fertilizer, total fertilizer and pesticides of respondent farmers in Chhoto-Maliha

Descriptions	Chemical Fertilizers (Rs.)	Bio Fertilizers (Rs.)	Total Fertilizers (Rs.)	Pesticides (Rs.)
Marginal	715.61	50.00	765.61	27.27
Small	675.85	57.48	756.35	21.74
Semi-Medium	761.12	63.25	829.10	13.06
Medium	760.89	78.89	839.06	24.75
Large	655.73	27.07	682.80	32.64
All Gr.	727.91	60.05	789.06	20.16

Source: Field Survey

Table 3.4. Levels of mechanization of the respondent farmers in the sample villages

Description	Nashigram	Kashiyara	Hapania	Chhoto-Maliha
No. of tractors used per acre	0.0452	0.0533	0.0294	0.0109
No. of threshers used per acre	0.0706	0.0738	0.0784	0.04957
No. of sprayers used per acre	0.0622	0.1189	0.0981	0.102226
No. of dusters used per acre	0.06496	0.0453	0.0588	0.0219
No. of submersibles used per acre	0.33079	0.0697	0.1472	0.077
Expenses on modern implements per acre (Rs.)	625.77	1165.41	322.35	331.23
Expenses on submersibles per acre (Rs.)	2000	1500	1750	2500

Source: Field Survey

Table 3.5. Average yearly net profit per acre of land from different Cropping systems followed by the respondent farmers in Nashigram

Cropping Systems	No. of Farmers	Land Size and Land in Acres	Gross Cropped Area in Acres	M.C.I.	Total Yearly Return (Rs.)	Total Yearly Production Cost(Rs.)	Average yearly Returns per Acre (in Rs.)	Average yearly Production Cost per Acre (in Rs.)	Average yearly Net Profit per Acre (in Rs.)
I	7 (14)	89.94 (24.19)	150.24	1.67	1775890	922782	19745.27	10259.97	9485.3
II	9 (18)	92.2 (24.79)	142.5	1.545	1631300	835331	17693.06	9059.99	8633.07
III	1 (2)	4.0 (1.07)	6	1.5	67454	30058	16863.5	7514.5	9349
IV	2 (4)	2.0 (0.54)	3.6	1.8	29830	12377	14915	6188.5	8726.5
V	2 (4)	10.34 (2.78)	2.08	1.942	247535	133875	23939.55	12947.29	10992.26
VI	21 (42)	127.42 (34.27)	218.32	1.713	2569447	1092400	20165.18	8573.22	11591.96
VII	2 (4)	10.7 (2.88)	10.7	1	116215	61610	10861.21	5757.94	5103.27
VIII	1 (2)	4.2 (1.13)	4.88	1.162	53270	28200	12683.33	6714.29	5969.05
IX	1 (2)	5.6 (1.51)	8.9	1.589	90061	58312	16082.32	10412.86	5669.46
X	2 (4)	3.8 (1.02)	4.8	1.263	51890	21585	13655.26	5680.26	7975

XI	1 (2)	1.6 (0.43)	1.72	1.075	20675	9656	12921.87	6035	6886.87
XII	1 (2)	20.0 (5.38)	34.4	1.72	429154	230680	21457.7	11534	9923.7
12	50 (100)	371.18 (100.0)	606.14	1.63	7082721	3436866	19049.81	9243.86	9805.96

Source: Field Survey. Figures in the brackets are in Percentages.
Gr.1=A.B.M.P.; Gr.II=A.B.M.; Gr.III=A.T.; Gr.IV=A.M.T.; Gr.V=A.B.T.; Gr.VI=A.B.; Gr.VII=A, Gr.VIII=A.M.P.W.; Gr.IX=A.B.M.W.; Gr.X=A.M.; Gr.XI=A.M.P.;Gr.XII=A.B.M.T.Pk.
A=Aman; B=Boro; W=Wheat; M=Mustard; T=Til; P=Potato; Pk=Pumpkin

Table 3.6. Average yearly net profit per acre of land from different cropping systems followed by the respondent farmers in Kashiyara

Cropping Systems	No.of Farmers	Land Size and Land in Acres	Gross Cropped Area in Acres	M.C.I.	Total Yearly Return (Rs.)	Total Yearly Production Cost(Rs.)	Average yearly Returns per Acre (in Rs.)	Average yearly Production Cost per Acre (in Rs.)	Average yearly Net Profit per Acre (in Rs.)
I	1 (2)	0.5 (0.22)	1	2	22905	15417	45810	30834	14976
II	13 (26)	46.06 (20.01)	105.84	2.298	2009899	1016964	43636.54	22079.11	21557.42
III	3 (6)	7.4 (3.21)	131.44	1.324	184958	110708	24994.32	14960.54	10033.78
IV	13 (26)	46.0 (19.99)	9.8	2.857	2510190	1997012	54569.35	43413.3	11156.04
V	12 (24)	89.0 (38.67)	196.3	2.206	3484480	2141142	39151.46	24057.77	15093.68
VI	5 (10)	26.0 (11.29)	62.2	2.392	1313476	737770	50518.31	28375.77	22142.54

VII	3 (16)	15.2 (6.60)	37.8	2.487	785583	441009	51683.09	29013.75	22669.34
07	50 (100)	230.16 (100)	544.38	2.365	10311491	6460022	44801.4	28067.53	16733.88

Source: Field Survey. Figures in the brackets are in Percentages.
Note for Table 3.6,
Gr-1=A.P.; Gr-1I=A.B.P.; Gr-III=A.B.P.M.; Gr-IV=A.B.P.T.;
Gr-V=A.B.P.M.T.; Gr-VI=A.B.P.T.Pk; Gr-VII=A.B.P.M.Pk.
A=Aman; B=Boro;M=Mustard;T= Til;P=Potato;Pk=Pumpkin.

Table 3.7. Average yearly net profit per acre of land from different cropping systems followed by the respondent farmers in Hapania

Cropping Systems	No. of Farmers	Land Size and Land in Acres	Gross Cropped Area in Acres	M.C.I.	Total Yearly Return (Rs)	Total Yearly Production Cost(Rs.)	Average yearly Returns per Acre (in Rs.)	Average yearly Production Cost per Acre (in Rs.)	Average yearly Net Profit per Acre (in Rs.)
I	5 (10)	5.8 (5.69)	10	1.724	99453	72719	17147.07	12537.76	4609.31
II	10 (20)	22.4 (21.97)	45.02	2.009	533487	351973	23816.38	15713.08	8103.3
III	13 (26)	36.8 (36.09)	78.9	2.144	1012223	600031	27506.06	16305.19	11200.87
IV	1 (2)	2.6 (2.55)	4.9	1.885	59045	38300	22709.62	14730.77	7978.85
V	1 (2)	6.0 (5.88)	12.5	2.083	137710	82935	22951.67	14730.77	9129.17
VI	2 (4)	4.2 (4.12)	8.3	1.976	87524	53610	20839.05	12764.29	8074.76
VII	1 (2)	2.0 (1.96)	4	2	34033	28362	17016.5	14181	2835.5
VIII	1 (2)	2.0 (1.96)	4	2	41839	26080	20919.5	13040	7879.5
IX	1 (2)	2.4 (2.35)	4.4	1.833	48367	29997	20152.92	12498.75	7654.17
X	1 (2)	2.0 (1.96)	6	3	56040	33155	28020	16577.5	11442.5
XI	1 (2)	4.0 (3.92)	8.2	2.05	108154	53725	27038.5	13431.25	13607.25
XII	1 (2)	0.76 (0.74)	1.52	2	17190	11260	22618.42	14815.79	7802.63
XIII	4 (8)	2.1 (2.06)	5.5	2.619	74750	49142	35595.24	23400.95	12194.29

XIV	1 (2)	0.4 (0.39)	1.2	3	12000	7460	30000	18650	11350
XV	1 (2)	0.6 (0.59)	1.2	2	16059	9850	26765	16416.67	10348.33
XVI	1 (2)	0.3 (0.29)	0.82	2.733	9888	4670	32960	15566.67	17393.33
XVII	1 (2)	1.6 (1.57)	3.2	2	39026	25097	24391.25	15685.63	8705.625
XVIII	1 (2)	1.6 (1.57)	3.7	2.313	35174	29628	21983.75	18517.5	3466.25
XIX	1 (2)	1.6 (1.57)	3.5	2.187	41975	27413	26234.38	17133.13	9101.25
XX	1 (2)	1.2 (1.17)	3.6	3	45420	25955	37850	21629.17	16220.83
XXI	1 (2)	1.2 (1.17)	2.8	2.333	31080	13220	25900	11016.67	14733.33
21	50 (100)	101.96 (100)	213.2	2.09	2540437	1574582	24916.02	15443.13	9472.88

Source: Field Survey. Figures in the brackets are in Percentages

Gr-I=A.B.;Gr-II=A.B.O.J.;Gr-III=A.B.O.J.Sp.;Gr-IV=A.B.O.J.Sp.Sc.;Gr-V=A.B.O.J.Sp.Sc.M.; Gr-VI=A.B.O.J.Sc.;Gr-VII=A.B.J.Sp;Gr-VIII=A.B.J.;Gr-IX=A.B.O.J.M.;Gr-X=A.B.M.; Gr-XI=A.B.O.J.M.Ca.; Gr-XII=B.J.Sp.; Gr-XIII=O.J.Sp.; Gr-XIV=A.O.J.; Gr-XV=A.B.J.Ch.Tu.; Gr-XVI=A.B.O.J.;Gr-XVII=A.B.O.J.Ch.; Gr-XVIII=A.B.O.J.Sp.P.M.; Gr-XIX=A.B.O.J.Sp.Ch.; Gr-XX=A.O.J.Sp.; Gr-XXI=A.J.M.Cu.

A=Aman; B=Boro; P=Potato; Sp=Sweet Potato; O=Onion; J=Jute; M=Mustard; S=Sugarcane; C=Chillies; Tu=Turnip; Ca=Cabbage; Cu=Cucumber.

Table 3.8. Average yearly net profit per acre of land from different cropping systems followed by the respondent farmers in Chhoto-Maliha

Cropping Systems	No. of Farmers	Land Size and Land in Acres	Gross Cropped Area in Acres	M.C.I.	Total Yearly Return (Rs)	Total Yearly Production Cost(Rs.)	Average yearly Returns per Acre (in Rs.)	Average yearly Production Cost per Acre (in Rs.)	Average yearly Net Profit per Acre (in Rs.)
I	1 (2)	24.0 (13.19)	33	1.375	347622	120916	14484.25	5038.17	9446.08
II	1 (2)	18.0 (9.89)	29.8	1.655	291189	107235	16177.17	5957.5	10219.67
III	1 (2)	4.8 (2.64)	8.8	1833	102184	43711	21288.33	9106.46	12181.88
IV	2 (4)	12.0 (6.59)	13.58	1.131	139364	49848	11613.67	4154	7459.67
V	10 (20)	35.0 (19.23)	51.9	1.483	512876	261686	14653.6	7476.74	7176.86
VI	1 (2)	7.2 (3.95)	8.3	1.153	85648	36530	11895.56	5073.61	6821.94
VII	1 (2)	1.2 (0.66)	2.4	2	17670	4430	14725	3691.67	11033.33

VIII	8 (16)	11.2 (6.15)	11.2	1	113506	47889	10134.46	4275.8	5858.66
IX	5 (10)	11.2 (6.15)	13.6	1.214	128619	58494	11483.84	5222.68	6261.16
X	2 (4)	1.6 (0.88)	3.2	2	27790	13238	17368.75	8273.75	9095
XI	6 (12)	14.8 (8.13)	28.54	1.928	280779	153438	18971.55	10367.43	8604.12
XII	5 (10)	11.62 (6.38)	18	1.552	143157	55254	12319.88	4755.08	7564.8
XIII	3 (6)	6.(3.4)	8.4	1.354	93926	38634	15149.35	6231.29	8918.06
XIV	1 (2)	2.8 (1.54)	3.7	1.321	28100	17490	10035.71	6246.43	3789.28
XV	1 (2)	9.6 (5.27)	15	1.562	149.361	73585	15558.44	7665.1	7893.33
XVI	1 (2)	6.0 (3.29)	7.9	1.317	67795	32872	11299.17	5478.67	5820.5
XVII	1 (2)	4.8 (2.64)	6.8	1.417	72868	26680	15180.83	5558.33	9622.5
17	50 (100)	182.0 (100)	264.12	1.451	2602454	1141930	14299.2	6274.34	8024.76

Source: Field Survey. Figures in brackets are in percentages.
Gr.1=A.B.M.P.Su.Ke;Gr.II=A.M.P.;Gr.III=A.B.M.T.P.Su.Ar.;Gr.IV=A.M.P.;Gr.V=A.B.M.; Gr.VI=A.M.P.Su.; Gr.VII=A.M.Kh; Gr.VIII=A.; Gr.IX=A.M.; Gr.X=A.M.T.P.;Gr.XI=A.B.M.P.; Gr.XII=A.M.T.; Gr.XIII=A.B.; Gr.XIV=A.M.Su.;Gr.XV=A.B.M.T.P.; Gr.XVI=A.M.T.P.Su.;Gr.XVII=A.B.M.T.Su.Br.Ar.
A=Aman; B=Boro; M=Mustard; T=Til; P=Potato; S=Sugarcane; B=Brinjal; Ar=Arum; Kh=Khesari.

Table 3.9. Overall Agricultural Performance of the respondent farmers in the sample villages (all crops taken together) for the Agricultural Year 2005–06

Description	Nashigram	Kashiyara	Hapania	Chhoto-Maliha
Average Physical Productivity (kgs.)	1,722	4,157	2,254	1,642
Average Effective Value Productivity (Rs.)	11,671	18,942	11,901	9,830
MCI	1.630	2.365	2.092	1.451
Average Production Cost Per Acre (Rs.)	9,244	28,067	15,443	6,274
Average Annual Net Profit Per Acre (in Rs.)	9,850	16,734	9,473	8,025
Benefit Cost Ratio (BCR)	1.065	0.596	0.613	1.279

Source: Field Survey

CHAPTER FOUR

UNDERSTANDING THE CONSEQUENCES OF LAND ACQUISITION AMONG FEMALE PEASANTS: A CASE STUDY FROM SOUTH WEST BENGAL

ARUP MAJUMDER

1. Introduction

Development-induced displacement is a contesting issue throughout the world. Deprivation of common people in the name of national growth is not a new thing, but on the other hand without national growth developmental activities cannot be effectively carried out. All development-oriented programmes are initiated for the sake of increased well-being of the entire nation, but in reality they are often observed to cater to the well-being of a certain section of the people, while a major part is affected by untoward intervention in their natural environment. Every year, more than fifteen million people are displaced globally due to the initiation and completion of several developmental projects (Mathur 2006). In India, fifty million people were displaced during the last fifty years after independence (Roy 1999).

This man-made forced displacement exposed the population to several risks of impoverishment, socially, psychologically and economically. They feel powerless when they are uprooted from their natural habitats and are compelled to accept the fact as fate. A small minority, mostly from the upper class, may improve themselves economically but most of the poor, Dalits and Tribals lose out with the intervention of development projects creating a divide between the "haves and have nots" (Mahapatra 1999). Tribal population in India constitutes 8.6% of the total population where it constitutes 40% of the DPs or PAPs (Fernandes 2008). Women also suffer a great deal. The deprived population is marginalized in terms of social and psychological status in addition to economic marginalization. This deprivation also prevents them from being aware of their personal as well as community strengths (Cernea 1997).

In this chapter a field-based anthropological study has been undertaken among a group of peasant families in a village under Kharagpur I Block in the Paschim Medinipur district, West Bengal, regarding the impact of acquisition among the women of land-losing families.

2. Area and the people

The villages of the study area come under the administrative jurisdiction of the Kharagpur I block in the Paschim Medinipur district. The Kharagpur I block is situated in the western part of the district and is bounded tothe north by the *Kasai* river. On the west and the south of the block lie the Jhargram sub-division, while the Kharagpur township is located to the east. Although the two major townships of the district are situated in the vicinity of this block, it is chiefly an agricultural area with few patches of *sal* forest. The area is characterized by vast open cultivable lands interspersed with village settlements connected by un-metalled roads.

According to a survey conducted by the Block Development office in 1997–98, the block has an area of 201 sq. km. or 27,979.21 hectares, within which 18,500 hectares are under cultivation (66.12%). The same survey has also found that about 6,905 hectares of the total cultivated area are under more than one crop, which turns out to be 37.32% of the cultivated land of the block. The total population of the block is 1,21,685, in which the males outnumber the females (male 62,314 and female 59,364), and there are 22,666 people belonging to scheduled caste (18.62%), while 29,974 (24.63%) belong to the scheduled tribes. Among the scheduled caste people the females (11,683) outnumber the males (10,983), while within the scheduled tribe population the gender ratio is in favour of the males (15,528 males, and 14,448 females). The population density of the block turns out to be a little more than 605 persons per sq. km. while the average household size is slightly above 5 persons. The district statistical handbook report based on the 1991 census data, however, differs from the survey conducted by the KharagpurI Block Development Office. According to the 1991 census figures, the total population of the block had been recorded as 1,21,659, while the area of the block was 281.94 sq. km. and this gave a population density of 432 persons per sq. km, with 268 mouzas of which 225 were inhabited (District Statistical Handbook, Medinipur 1998).

3. Methodology

The study mainly depends on direct intensive observation, interviews and the collection of case studies from the villagers affected by land acquisition. The survey was conducted among the households of the Gokulpur village with the help of structured and open-ended questionnaires. The qualitative information regarding the feelings and attitudes of the persons were collected through repeated conversations with them over a long period from the 210 female members of land-losing families and 512 female members among the non-land losing families.

4. Dependence of the villagers on agriculture

The name of the village specifically studied is Gokulpur, which is a multi-ethnic farming village situated about 7 km from the town of Medinipur. This village is located very near the river *Kasai* in the east, and in the west lies the south-eastern railway track which runs between Medinipur and Kharagpur railway stations.

On the east of Gokulpur lies the village Borkola. On the south side of this village are Chunpara and Nimpura. In the west and north lie the villages Amba and Ajobpur, respectively. Except for Chunpara, all the other villages that surround Gokulpur are agricultural villages in which most of the inhabitants depend on agriculture and its related economic pursuits. In Gokulpur, the majority of the villagers are also dependent on the cultivation of paddy and various kind of vegetables.

There are two main types of land in Gokulpur termed *Jal Jami* and *Kala Jami* in local parlance. The former are low lying and hold water during the rainy season, while the latter are located near the house-sites and at higher elevations. In the rainy season, people mainly cultivate paddy in the *Jal Jami*, while *Kala Jami* is used for vegetable cultivation in the winter season. In Gokulpur, 48.26% of land belongs to *Jal Jami*. This has been calculated from the data collected by our household census survey.

Table 4.1. Land type in Gokulpur

Land Type (acres)		**Total**
Kala Jami	***Jal Jami***	
61.89 (51.74%)	57.73 (48.26%)	119.62

The following table describes the land holding pattern of the households of Gokulpur, constructed on the basis of the two types of land owned by the villages.

Table 4.2. Land holding pattern of the village before acquisition

Size category of landholding in acres	Number of households	Mean household size
Landless	98 (25.26%)	4.50
≤ 0.5	59 (15.21%)	4.38
0.5–1.5	87 (22.42%)	4.85
1.5–2.5	61 (15.72%)	4.15
2.5–3.5	27 (06.96%)	4.55
3.5–4.5	28 (07.22%)	7.28
4.5–5.0	10 (02.58%)	9.90
5.0 +	18 (04.64%)	6.67
Total	388	4.95

This table reveals the nature of landholding patterns in Gokulpur before the acquisition of land by the Tata Metaliks Company. Land is one of the most vital life support resources for the peasant families of this area, but that does not mean that all the families of this village owned agricultural land before the acquisition. About 25% of families in the village were landless even during the pre-acquisition period; the remaining 75% owned some amount of agricultural land. The landholding pattern of these families however shows that there were fewer families (about 20%) who owned 2.5 to 5.0 or more acres of land. The majority of families in this village (about 53.35%) owned 0.5 to 2.5 acres of land. The highest percentage of families in a single category of land holding belongs to the cohort owning 0.5 to 1.5 acres of land, and this is 22.42% of the total number of families.

We should also look into the family size *vis-a-vis* the land ownership pattern in the village to understand the nature of dependency on land. The mean household size of the families and their corresponding landholding categories clearly reveal a pattern, showing that families ranging from the landless category up to the level owning 3.50 acres of land supported 4.50 persons on average, which is also very close to the average household size of the total village. However, the average household size rises to more than seven persons as soon as the level of landholding crosses 3.50 acres and the former rises to nine people with a landholding between 4.5 to 5.0 acres. This relationship between household size and landholding pattern implies that it is agricultural land which supported a higher number of people in a family. The earlier study based on field surveys done in Gokulpur during 1996 (after land acquisition) also revealed the preponderance of landless and land-poor families, showing 34% for landless families and about 49.40% for landholding (≤ 0.5–1.5 acres) out of the 329 families (Guha 2007). Comparing the two surveys it can be said that although there is no radical shift in the landholding pattern in the village over the twelve years, some improvement has however occurred in terms of landownership by the families in this village. For example, in 1996, there were fourteen families owning 2.5 to 3.5 acres of land, while according to the 2007 survey 27 families (out of 388) were found to belong to this landholding category. On the higher side, the situation is more revealing. There were only seven families owning between 3.5 to 7.5 acres of land in 1996, whereas in 2007 as many as 56 families belonged to the above landholding category. Does this mean that during 1996 to 2007 some land was distributed to the villagers of Gokulpur through land reform? The household census and economic surveys conducted during 2007 in the village did not yield any case of land distribution by the Land Reforms Department of the Government. Our field interviews revealed that many families of Gokulpur purchased agricultural land and some received land as dowry, which is a very common practice among the peasant families in this area. It seems that we have to investigate in more detail regarding the improvement in the landowning scenario in Gokulpur, which of course is not related directly to acquisition of land for the industry. However, the improvement in the landownership pattern proves another point—although the peasant families in Gokulpur are living under a constant threat of land acquisition because of the arrival of many industries in this area, they are still trying to stick to their agricultural occupation through landownership.

5. Land acquisition for Tata Metaliks

Tata Metaliks is a heavy industry established within the jurisdiction of Kalaikunda Gram panchayat during 1992. This is a pig-iron manufacturing plant, found to produce about 290 tonnes per day in 1995–96. After the establishment of Tata Metaliks the company has built up a metal road on the western side connecting the plant with national highway 6 in Sahachawk. The south-eastern railway line runs on the eastern side of the industry. The Kharagpur railway station is only about five kilometres and the Medinipur district headquarters is seven kilometres from this place. Connected to this, on June 1, 1992, the Land Reforms Ministry mentioned that 217.23 acres land were acquired for the Tata Metaliks (Guha 2007, 85).

The land acquired for the pig-iron industry belonged to the *jal some* class according to the age-old system of classification made by the Land and Land Reforms Department. The possessions on these lands were given to the company on different dates in August 1991, and declaration notifications were published from November 1991 to January 1992. The Land Acquisition Department approved a rate of Rs. 20,686 per acre. The cases of Land acquisition for Tata Metaliks have shown that the Government of West Bengal desired a quick acquisition of land for the company, and that is why Act II was employed for the said purpose (Ibid., 87).

In Table 4.3 below we have shown the agricultural landholding scenario of the land-losing families in the village both before and after the acquisition.

Table 4.3. Pre-acquisition and post-acquisition agricultural landholding scenario of the land-losing families in the village

Size category of landholdings in acres	Before land acquisition	Mean household size	After land acquisition	Mean household size
Land less	-	-	5 (5.05%)	3.60
≤ 0.5	10 (10.10%)	1.90	28 (28.28%)	4.21
0.5–1.5	38 (38.38%)	3.97	39 (39.39%)	6.21
1.5–2.5	23 (23.23%)	6.80	19 (19.19%)	5.16
2.5–3.5	17 (17.17%)	7.80	03 (03.03%)	9.66
3.5–4.5	04 (4.04%)	8.20	03 (03.03%)	9.33
4.5–5.0	-	-	1 (01.01%)	14.00
5.0 +	07 (7.07%)	9.40	1 (01.01%)	11.00
Total	99	5.63	99	5.63

This table shows the pre-acquisition and post-acquisition agricultural land holding scenarios of the land-losing families in Gokulpur. After land acquisition, five families among all 99 became landless, which constitutes 5.05% of the total land-losing families. The families with less than 0.5 acres of land dramatically increased from 10 (10.10%) to 28 (28.28%) after land acquisition. It was noticed that even after land acquisition the number of families with 0.5–1.5 acres of land increased from 38 (38.38%) to 39 (39.39%). However, the families with 1.5–2.5 acres of land decreased from 23 (23.23%) to 19 (19.19%). Similarly, a large decline was noticed in the number of families with 2.5 acres to more than 5.0 acres of land. The number of such families decreased from 28 (28.28%) to 8 (8.08%) after acquisition.

6. Impacts of land acquisition on women

In every agricultural as well as rural society the women usually possess a safe and secure place within the household premises. However, a sudden drastic fall in the economic level can force them towards the outer world in search of betterment. This betterment further enhances their social status at the intra family, inter family as well as at the social level.

7. Changing occupational patterns of women

The sudden loss of agricultural land due to acquisition leads to the economic impoverishment of the land-losing families that further motivates the women to step outside in search of a sustained source of livelihood.

Fig. 4.1. Changing occupational pattern of the land-losing female population after acquisition

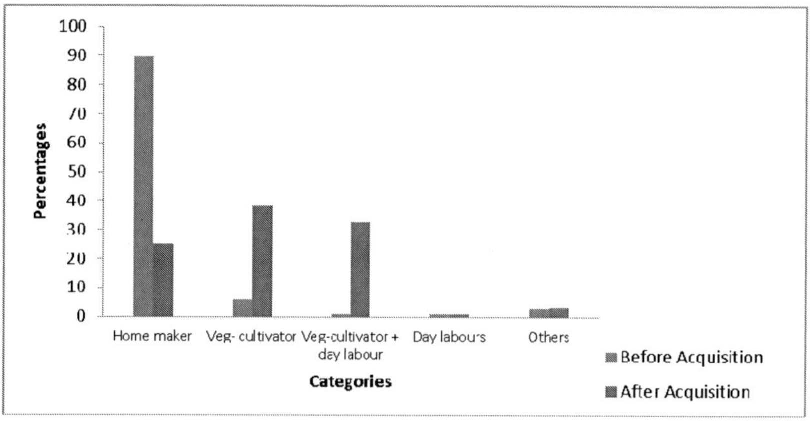

Before the acquisition of land, a very small number of women were engaged in earning a livelihood. Almost 90% of the total female population was homemakers, but after the period of acquisition it dropped to 25%. This sudden drop further led to crowding in the vegetable vending

and cultivator category of labour (see Fig. 4.1 above), which symbolizes that women were forced to step out of their homes in search of a livelihood which might contribute to their family expenses. They preferred to work as vegetable cultivators and sellers, as well as agricultural day labourers in the neighbouring areas. This economic enhancement after acquisitions not only helped these people to contribute and strengthen their family incomes, but also provided a good platform for decision making.

Fig. 4.2. The changing occupational pattern of the landowner/non-land-losing female population after acquisition

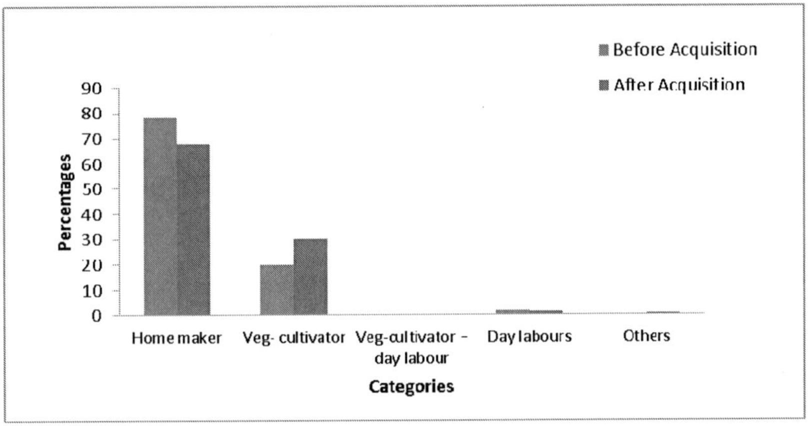

As the crunch of economic necessity was not related to the population who did not lose any of their land, forced occupational change was almost absent among them. Before acquisition about 80% of the population were homemakers, among whom only about 10% transferred themselves to the category of vegetable cultivators. The affected female community in this class cast a demonstration effect on the previously mentioned segment in motivating them to change their occupational pattern. Women who changed their occupations were eager to enjoy financial freedom like the affected women. Thus, no drastic change can be observed throughout. Two interesting facts can be observed in this aspect: they did not maintain two jobs at the same time and did not prefer working as day labourers in both the pre- and post-acquisition periods. This bears testimony to the fact that they were neither in "urgency" nor in "necessity."

Case 1

Archana Ghose, a sadgope housewife, earns her livelihoodby cultivating vegetables in herhomestead area. After two months of marriage her in laws lost 3 acres of land. At that time there were seven members who depended on the same hearth. After the acquisition they were left with 0.5acres of land, which was not sufficient to feed the whole family. She was therefore advised by her husband to work and earn for the family,but being a newly married bride it was quite impossible for her to do so at that time. At thesame time her mother-in-law planned to sell vegetables that they grew in the kitchen garden area. She herself started assisting her mother-in-law along with her three sisters-in-law and soon they started selling in the Kharida Bajar, which is situated in Kharagpur. At present she is the lady of her family and a mother of two. Her elder son left school two years ago and shestill feels sorry for this, as she did not get a chance to guide him or support him through spending a lot of time outside the house. She believes that if she hadmore time to interact with her son this would not have happened.

8. Utilization of money earned by women

The main objective behind earning money in times of necessity is generally related to survival, while in other cases it is mostly related to economic freedom. The degree of necessity and utilization of earned money is completely dependent on the prevalent financial status of the earning persons.

The women utilized their incomes in several ways. The majority of the land-losing women, as well as the land-owners, share the common intent to deposit a part of it in the bank for the future. Again, land losers also paid much attention to repaying their loans (37.63%), but the majority of the land owners liked to spend it on household purposes besides bank deposition. A multiple mode of consumption is also common, which includes expenditure on educational purposes. Here, the land owners enjoy a better position than the losers, as it is very obvious that people who want to earn to enjoy financial freedom spend more on extra-necessary consumption than the people who work for their bare survival.

Fig. 4.3. Utilization pattern of cash income by women

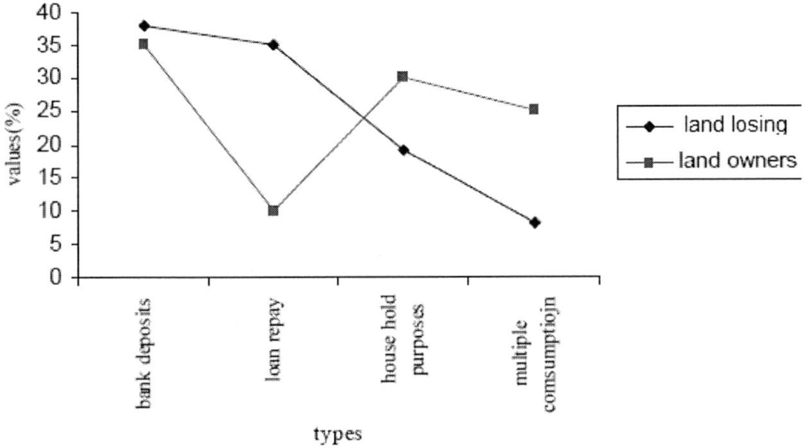

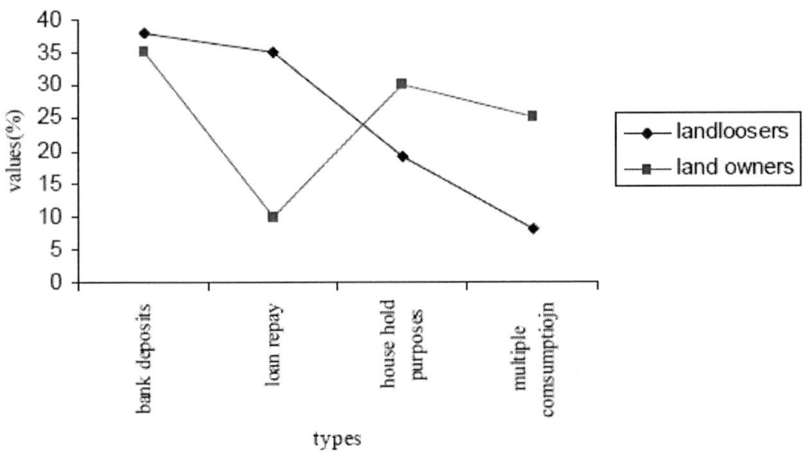

9. Changing pattern of school dropouts after acquisition

In rural areas there is a general tendency for increased school dropouts, as the people are not fully aware of the importance of education. In addition, the misbalanced economic life can worsen the whole situation and accelerate the dropout rate. As the study objective is to measure the dropout rate along with focusing on its transition over two decades, a back calculation has been made to trace the trend of dropouts.

In order to measure the dropout trend, the 6–16 years age range is generally used. However, as per the requirement of the study objective, a back calculation has been done to ensure accurate results. Thus, in this case the age range is 6–37 years. For example, if a boy was 21 during the period of acquisition and lost his scope of further study due to financial pressure, then he should be considered a dropout. At present, his age would be 37, but in this study he is still considered a dropout and chosen as a sample member of the whole dropout population to understand the entire scenario.

10. Comparison of school dropout status

At the time of our fieldwork we found that although the educational achievements of the women in Gokulpur were lower than those of the males, more women of the younger age groups were attaining primary and upper primary levels of school education. Under this general trend of educational achievement we have collected data through interviews and case studies on the educational scenario of the land-losing families in the village. One of the most adverse effects of land loss was reflected in the discontinuation of education of the children by the parents. It was found that female children were most deprived in this process of dropping out of school education. Table 4.4 below shows a comparison of the dropout of female children among the land-losing and non-land-losing families from primary, upper primary and secondary levels of education. The table shows that irrespective of the level of education, the number of females who discontinued school education is higher among the land-losing families than the non-land-losing families. The comparative scenario is stark at the primary level, in which more than 26% of the enrolled girls dropped out of school after the acquisition of land, while there was not a single dropout case among the non-land-losing families, which means that the parents of non-land-losing families continued the education of their female children at the primary level. The comparative figures however showed that a sizeable section of girls dropped out of school education at the upper primary and secondary levels both among the land-losing and non-land-losing families, but the percentage of dropouts among the land-losing families was much higher than that among the latter. The overall picture shows that land acquisition caused a greater number of dropouts among girls in Gokulpur.

Table 4.4. Distribution of female dropouts among land-losing and non-land-losing families

Educational level	Landlosing		Non-landlosing	
	No. of dropouts	Total no. of enrolled girl children	No. of dropouts	Total no. of enrolled girl children
Primary	7 (26.92%)	26	-	83
Upper primary	35 (67.30%)	52	26 (35.13%)	74
Secondary	27 (93.10%)	29	35 (52.23%)	67
Total	69 (64.48%)	107	61 (27.23%)	224

11. Declining age at marriage

Age at marriage is regarded as one of the important parameters for assessing the degree of freedom enjoyed by the women in particular societies, and is often considered one of the essential socio-demographic features for measuring the degree of modernization in a society. In traditional societies women were usually married off at a much lower age, which gave them very little opportunity to obtain an education and forced them to undergo the hardships of childbearing and motherhood from a very early age. With advancement of societies, the scope of female education increased and the women also began to compete with men in the job market. The broadening of economic and educational opportunities, changes in the value systems, various social reform measures and governmental legislations gradually caused an upward trend in the age of marriage among women both in the rural and urban areas of India. Against this general background we have made an attempt to observe whether land acquisition, which caused dropouts of female children from school education, had any effect on the age at marriage or not. In order to look into the relationship between age at marriage and land acquisition we have collected data on the age at marriage of those women who were married off by their parents within two years after acquisition. Table 4.5 below compares the number of girls among the land-losing and non-land-losing families at different categories of age at marriage. It was interesting to see that the percentages of women married below the constitutional age were much higher among the land-losing than the non-land-losing families. Not only this, in almost each category of age, excepting 20–23, the proportions of females married were higher in land-losing than that in non-land-losing families. We enquired about the matter in the field and found that one of the reasons that women were married off among the land-losing families at lower ages was due to their parents being anxious about their futures, and that the compensation money usually spent to pay the dowry in a daughter's marriage was rather high. At higher age groups, girls in non-land-losing families were usually allowed to continue their education or do some work in the family. Hence, the proportion of married girls in higher age groups, excepting some specific cases, were comparatively less in such families than in land-losing families where grown-up girls are usually considered a burden and preferred to be married in order to gain some financial relief. A typical case study, described below, reveals the realities on the ground.

Table 4.5. Age at marriage of girls of the land-losing and non-land-losing families

Age categories (in years)	Landlosing		Non-landlosing	
	No. of girls married off	Total no. of daughters	No. of girls married off	Total no. of daughters
Below 18	7 (47.05%)	10	3 (23.08%)	13
18–20	3 (42.85%)	7	9 (33.33%)	27
20–23	4 (80%)	5	7 (87.5%)	8
23–26	-	2	2 (22.22%)	9
26–29	2 (40%)	5	2 (25.00%)	8
29–32	1 (33.33%)	3	1 (14.28%)	7
Total	17	32	24	71

Case 2

Bishnu Santra, a Sadgope farmer, lost 2.75 acres of land, being left with 2 acres and four dependents. Unlike other farmers he was affected by the acquisition, and although the amount left was not sufficient they somehow managed. Immediately after the acquisition, his wife started working as a day labourer along with him. After six months they decided to look for a well settled groom for his eldest daughter who was just fifteen at that time. However, the received proposals were not satisfactory in terms of social status. He said "*jon khatche bole loke mone korto amra duley bagdi*," which means that people thought that they belonged to a relatively lower class than their actual caste. Here, the phrase *duley bagdi*refers to the lowest class of people who are generally appointed as day labourers in the agricultural fields of Sadgopes. He was apparently disappointed, but later understood that it was not their fault. He had to pay theamount of Rs. 25,000 in cash as dowry to get a suitable groom of the Sadgope caste. He used the entire amount of compensation for the marriage of his eldest daughter and he isnow left with two more. He still doesn't know how he will manage for them.

12. Concluding remarks

This study emphasizes how land acquisition is forcing changes on the status of women in the study area. The major finding shows that land acquisition in a rural area not only leads to landlessness, but also, like the male members, female members of the land-losing families have been affected indifferent ways. Besides this, our data showed that the school dropout rates or out of school status among female members of land-losing families have increased more than that in non-land-losing families. This study is also reflective of the fact that after the acquisition, livelihood patterns have changed among the female members of land-losing families, and the "age at marriage" has decreased more on average among the girls of land-losing families than in non-land-losing families.

References

Cernea, M. *Socio-Economic and Cultural Approaches to Involuntary Population Resettlement.* World Bank, Oxford: Oxford University Press, 1991.

Guha, A. "Giving Tiller's Land to The Capitalists: An Ethnography of Land Acquisition in West Bengal." *Journal of the Indian Anthropological Society* 41 (2006): 223–241.

—. *Land, Law and The Left: The Saga of Disempowerment of the Peasantry in the Era of Globalization.* New Delhi: Concept Publishing Company, 2007.

Majumder, A. "Changing in Socio-Economic Condition of Landloser Families in a Village in Paschim Midnapore: An Anthropological Assessment of Development Caused Displacement." Department of Anthropology, Vidyasagar University, Medinipur (Unpublished M.Sc. dissertation), 2007.

Majumder, A. & A. Guha. "A Decade after Land Acquisition in Paschim Medinipur, West Bengal." *Journal of the Indian Anthropological Society* 43 (2008): 121–133.

Majumder, A. & A. Guha. "The Impact of Land Acquisition on the Joint-Extended Family among the Peasants of Gokulpur, Passchim Medinipur." *Indian Antropological Society* 44 (2009): 69–75.

SECTION B:

GENDER ISSUES

CHAPTER FIVE

MICROFINANCE ACCESS AND FEMALE EMPOWERMENT IN INDIA: AN INTER-STATE ANALYSIS

ARINDAM LAHA AND PRAVAT KUMAR KURI

1. Introduction

In the context of developing countries, a well designed microfinance programme by SHG – bank linkage model is considered to be an effective institutional mechanism to promote socio-economic development of the economically weaker sections of the society especially to the women by breaking the prevailing credit constraints. Generally, women face difficulty in accessing institutional credit due to lack of assets that can be used as collateral. Microfinance as an alternative delivery mechanism targets women because women are more likely to be credit constrained than men (Pitt and Khandker, 1998). The programme has its inherent capacity to unveil the untapped potentiality of women by mobilizing them to pool their own funds, building their capacities, and empowering them to leverage external credit (Zubair, 2006). Microfinance programmes in Asia and Pacific countries are found to be successful in extending the services to the poorest and women sections of the population. In these continents, nearly 62 percent of the members covered under such programmes are considered as poor women (Maes and Reed, 2012). Historically, Bangladesh dominates among the countries of South Asia in terms of microfinance outreach and share of total borrowers. The Grammen Bank of Bangladesh is a well known people's institution for the poor women promoting small scale production by extending microcredit. By the end of 2011, it had served over 8 million borrowers, of whom 96 percent were women. Taking the lessons from Bangladesh, micro finance programme has been emerging in a big way in India. In fact, India has achieved a considerable success in expanding the outreach of micro finance institutions to promote socio-economic development of the poor women. According to the recent statistics published by NABARD, around 80 per cent of the SHGs were exclusively formed by women and their share in the disbursement of loan is around 85.5 percent (NABARD, 2012).

Microfinance outreach has its inherent impact on the empowerment of women. Empowerment is considered as a multidimensional "process of increasing the capacity of individuals or groups to make choices and to transform those choices into desired actions and outcomes" (World Bank, 2002). Malhotra et al (2002) constructed a list of the most commonly used dimensions of women's empowerment, drawing from the frameworks developed by various authors in different fields of social sciences. In the framework, the process of women empowerment will simultaneously enlarge several spaces, e.g., economic, financial, socio-cultural, legal, political, psychological and other related spaces. An expansion of several dimensions of women empowerment will unfold a new horizon in the path of equitable development if expansion of one dimension is not achieved at the cost of another. In such multi-dimensional aspects, women's economic space is found to be central to alleviate poverty (Burra et al, 2005). To ensure a sustainable approach to women empowerment, it is desirable that microfinance programme is preceded by a process of social mobilization. A necessary intervention in the process of social mobilization can be visualized by the drive of women literacy programme.[1] Under this backdrop, exploring the present state of the spread of microfinance outreach across the states of India, this paper attempts to analyse the association between microfinance outreach and women empowerment in India. The inter-state variations in the outreach of microfinance and the extent of women empowerment have been measured by constructing two comprehensive indices viz. the index of microfinance outreach encompassing penetration, availability and usage components of microfinance, and the index of women empowerment by considering economic, financial and social dimensions of empowerment. For convenience, the paper is divided into six sections. The next section analyses the trends of outreach of microfinance programme in India overtime and participation of women in such programme. Section 3 considers the data and methodological aspects relating to the construction of indices on microfinance outreach and women empowerment. Section 4 examines the interstate variation in the level of microfinance outreach across states of India. The extent of women empowerment across states of

[1] Hirway and Mahadevia (1996) argued that education is a means to empowerment for women in the developing countries.

India is also presented in this section. Section 5 examines the association between microfinance outreach and women empowerment in the light of empirical evidences. The concluding remarks and policy implications appear in section 6.

2. Trends of microfinance outreach in India

SHG-Bank Linkage Programme (SBLP) has emerged as the major microfinance programme, even if there are other different models for extending microfinance in the country. The programme is being implemented by the commercial bank, RRBs and cooperative banks. The cumulative progress of SBLP has been shown in table 5.1. The performance of the programme in outreaching its services has significantly scaled up subsequently during the period of study. The growth rate in the disbursement and outstanding of loan are observed to be positive in all time periods. However, a deceleration trend is observed in the formation of new SHGs financed by the bank from 2009-10. The number of new SHGs financed by banks decreased from 15, 87,000 as on 2009-10 as against 11, 96,000 SHGs in 2010-11, registering a decline of around 24.64 percent. Moreover, 47.87 lakh SHGs had outstanding bank loans of 31,221 crore in 2011, as against 48.5 lakh SHGs with bank loans of `28,038 crore as on 31 March 2010. This represents a decline of 1.3 per cent in the number of SHGs and a growth of 11.4 per cent in bank loans outstanding to SHGs (Economic Survey, 2011-12). The programme of SBLP is specially designed to cater the needs of women section of the population. It has been seen that around 80 per cent of the SHGs linked were exclusive women SHGs and their share in the disbursement of loan is around 85.73 percent (Table 5.2).

Table 5.1. Progress under SHG-Bank Linkage

Year	New SHGs Financed by Banks During the Year			Bank Loan** Outstanding as on 31 March 2011		
	No.(lakh)	Amount (Rs.crore)	Growt h (%)	No.(lakh)	Amount (Rs.crore)	Growth (%)
2002-03	2,55,882	1,022.34	87.00	7,17,360	2,048.68	
2003-04	3,61,731	1,855.53	81.50	10,79,091	3,904.21	90.57
2004-05	5,39,365	2,994.25	61.37	16,18,456	6,898.46	76.69
2005-06	6,20,109	4,499.09	50.26	22,38,565	11,397.55	65.22
2006-07	11,05,749*	6570.39	--	28,94,000 @	12,366.49	--
2007-08	12,27,770*	8849.26	34.68	36,26,000	16,999.90	37.47
2008-09	16,09,586*	12,256.51	38.50	42,24,000	22,679.85	33.41
2009-10	15,87,000	14,453.30	17.92	48,52,000	28,038.28	23.62
2010-11	11,96,000	14,547.73	0.65	47,87,000	31221.17	11.35

Source : National Bank for Agriculture and Rural Development (NABARD), as mentioned in Economic Survey, 2011-12.
Note : * Include existing SHGs also, which were provided repeat bank loan.
** Includes repeat loans to existing SHGs.
@ from 2006-07 onwards, data in respect of number of SHGs financed by banks and bank loans are inclusive of SHGs financed under the Swarnajayanti Gram Swarozgar Yojana (SGSY) and the existing groups receiving repeat loans. Owing to this change, NABARD discontinued compilation of data on cumulative basis from 2006-07. As such data from 2006-07 onwards are not comparable with the dataof the previous years.

Table 5.2. Progress of exclusive women SHGs

Up to end March	Exclusive Women SHGs	Percentage of total SHGS	Loan disbursed	Percentage of total loan disbursed
2007-08	3986093	79.57	7474.26	84.46
2008-09	4863921 (22.0)	79.46	10527.38 (40.8)	85.91
2009-10	5310436 (9.2)	76.4	12429.37 (18.1)	86
2010-11	6098000 (14.8)	81.7	12622.33 (1.6)	86.8
2011-12	6299000 (3.3)	79.1	14132.02 (12.0)	85.5

Source: Status of Microfinance in India (various years)

3. Data and Methodology

Existing literature on the trends and progress of microfinance centers on the supply side perspectives from the point of view of bankers, NGOs and governments (NABARD, 2008; Sangwan, 2008; Verman, 2005). However, mere supply side solution does not provide a comprehensive picture of microfinance outreach. There is a need to take into account the demand side dimensions (NABARD, 2008). The demand for microfinance services is measured by means of actual utilization of the credit and savings services of microfinance programme, while the supply of microfinance services is measured by means of penetration and availability of credit services to the population. However, these indicators of financial access provide only partial information on the inclusiveness of the microfinance system of an economy. Due to some inherent limitations, these indicators fail to capture the overall extent of microfinance outreach adequately. In this regard, Srinivasan (2008) at first constructed a Microfinance Penetration Index (MPI) to capture the client mobilization effort in the context of its significance to the states' population. The index of penetration of microfinance was computed by the dividing the share of the state in microfinance clients by the share of the state in population. But, this index is based only on one dimension, viz. penetration of microfinance. An attempt has been made in the paper to construct a comprehensive measure of microfinance outreach that will capture information on several indicators of microfinance outreach viz. penetration, availability and usage. For each indicator, the performance of the state is evaluated in respect of the national average. For instance, to measure penetration of microfinance, at first, the share of number of SHG members[1] of a state to the country's total number of SHG members is computed. The indicator of microfinance penetration is then worked out as a ratio between the share of state in SHG members and the share of total population of the state. The score higher than one indicates higher penetration of SHG members' vis-à-vis proportion of total population in the state. In other words, the larger the distance from one greater is the SHG outreach in the state. In addition, the level of women's empowerment across the states of India is also measured in the study. The concept empowerment of women in its true sense follows a multi-dimensional approach. It encompasses economic, financial, social, political, legal and also sociological, psychological factors[2]. In our study, we have considered only three such dimensions like economic, financial and social empowerment to construct women empowerment index. Broadly, the construction of women's empowerment index follows the same methodology as IMO. For each dimension (economic, financial and social), the relative performance of the state in respect of national average is worked out. Then the relative performance of the state is divided by the share of the women households of the state to calculate relative share. For a clear exposition, the description of indicators used in the construction of the index of microfinance outreach and the index of women empowerment is given in the table 5.3.

[1] The study considers the total number of SHG members liked with public sector commercial bank, private sector commercial bank, regional rural bank and cooperative bank.

[2] Several attempts have been made in the literature to construct a comprehensive index encompassing several dimensions of women empowerment (Hirway and Mahadevia, 1996; Mehta, 1996).

Table 5.3. Description of the indicators and its data sources

Indicator	Description	Proportional measure	Data source
Indicator of Microfinance Penetration (MP_1)	Share of SHG members of the state as a proportion of the share of population of the state	MP_1=A/B Where A indicates the ratio of number of SHG members of the state to total number of SHG members in India, and B indicates the ratio of number of total population of the state to total number of population in India	Status of Microfinance in India (NABARD, 2012) and Census, 2011
Indicator of Microfinance Availability (MP_2)	Share of credit SHGs of the state as a proportion of share of total number of SHGs (savings and credit) of the state	MP_2=C/D Where C indicates the ratio of number of credit SHGs of the state to total credit SHGs in India, and D indicates the ratio of number of total SHGs of the state to total SHGs in India	Status of Microfinance in India (NABARD, 2012)
Indicator of Microfinance Usage (MP_3)	Share of volume of microfinance credit and saving as a proportion of share of NSDP of the state	MP_3=E/F Where E indicates the ratio of volume of saving and credit of the state to total volume of saving and credit in India, and F indicates the ratio of NSDP of the state to total NSDP in India	Status of Microfinance in India (NABARD, 2012) and CSO, 2011
Economic Empowerment	Share of women SHG	WE_1=G/H Where G indicates	Status of Microfinance

of women (WE_1)	members with banks (public, private, RRBs and Co-operatives) as a proportion of the share of women households of the state	the ratio of number of women SHG members of the state to total women SHG members in India, and H indicates number of women households of the state to total women households in India	in India (NABARD, 2012) and Census, 2011
Financial Empowerment of women (WE_2)	Share of bank account of women of the state as a proportion of share of women households of the state	WE_2=I/H Where I indicates the ratio of the number of bank account of women of the state to total bank account of women in India	Basic Statistical Return (RBI, 2011) and Census, 2011
Social Empowerment of women (WE_3)	Share of women literates of the state as a proportion of share of women households of the state	WE_3=J/H Where J indicates the ratio of the number of women literates of the state to total women literates in India	Census, 2011

To derive comprehensive index on microfinance outreach (IMO) women empowerment (IWE), we have used data driven weighting system derived from Principal Component Analysis. The comprehensive indices can be written as

$$IMO_s = \sum_{i=1}^{3} w_i MP_{is} \Big/ \sum_{i=1}^{3} w_i \quad \text{and} \quad IWE_s = \sum_{i=1}^{3} w_i WE_{is} \Big/ \sum_{i=1}^{3} w_i$$

where w_i (i=1, 2, 3) are the corresponding weights of the indicators. Depending on the values of IMO (IWE), states are categorized into three categories. States with an IMO (IWE) value below 1 are considered to have a low level of microfinance outreach (women empowerment), those in between 1 to 2 a medium level and those above 2 a high level.

4. Microfinance access and women empowerment across Indian states

In the study, the depth of microfinance outreach and extent of women empowerment in India is examined by constructing two composite indices of microfinance outreach and women empowerment. The index on microfinance outreach is based on the demand for and the supply of microfinance services. The multi-dimensional aspects of women empowerment have been conceptualized by considering three broad socio-economic indicators: economic, financial and social empowerment of women. The values of each dimension of these two indices across the states of India along with their ranks are presented in Table 5.4.

A wide inter-state disparity is visible in the individual indicators of microfinance outreach. In respect of the penetration of microfinance services, states like Kerala, Puducherry, Andhra Pradesh Tamil Nadu and Andaman & Nicobar Islands belong to the category of higher outreach of microfinance programme (as shown in column 2 of Table 5.4). These four southern states excel in outreaching microfinance programme to the financially excluded sections of the population. In fact, the shares of SHG members in those states are observed to be on average two times in comparison to the share of total population. On the other hand, a majority of nearly 84 percent of the states fall in the category of lower penetration of microfinance programme. The states in the lower end of the tail are mainly confined to the north-eastern, central, northern and eastern region. Inter-state disparity in the availability of microcredit is represented in column 3 of Table 5.4. Access to credit without any resort to collateral is considered as the most important financial service of the microfinance programme. It is evident that in Tripura, the share of credit SHGs is found to be doubled than the share of total number of SHGs working in state. In another six states (Goa, Andhra Pradesh, Puducherry, Tamil Nadu, Jammu & Kashmir and West Bengal), SHGs are more sustainable in having access of credit. In more than 77 percent of the states, most of the SHGs are now performing as a saving institution. However, actual utilization of microfinance services considers both savings and credit facilities. The usage of these facilities by the SHG members subject to the constraint of

Table 5.4. Ranking of the states on the basis of indicators of microfinance outreach and women empowerment (2011-12)

State/ Union Territories	MP_1	MP_2	MP_3	IMO	WE_1	WE_2	WE_3	IWE
Andaman & Nicobar Islands	2.2 (5)	0.9 (13)	0.44 (15)	1.12 (8)	1.93 (5)	1.07 (13)	1.28 (6)	1.4 (8)
Andhra Pradesh	2.54 (3)	1.6 (3)	5.04 (1)	3.17 (1)	3.01 (2)	1.46 (8)	0.95 (22)	1.73 (4)
Arunachal Pradesh	0.75 (15)	0.12 (31)	0.12 (27)	0.31 (30)	0.52 (18)	0.74 (21)	0.89 (25)	0.73 (23)
Assam	1.16 (10)	0.73 (17)	0.78 (8)	0.88 (10)	1.05 (11)	0.61 (24)	1.01 (20)	0.89 (16)
Bihar	0.41 (25)	0.9 (12)	0.75 (9)	0.7 (12)	0.4 (23)	0.42 (29)	0.77 (30)	0.54 (30)
Chhattisgarh	0.75 (16)	0.57 (23)	0.4 (16)	0.56 (20)	0.53 (17)	0.55 (27)	0.92 (23)	0.68 (26)
Goa	0.87 (14)	1.71 (2)	0.29 (22)	0.93 (9)	0.89 (14)	5.27 (1)	1.3 (3)	2.49 (1)
Gujarat	0.49 (22)	0.94 (11)	0.16 (25)	0.52 (21)	0.44 (20)	1.05 (15)	1.09 (15)	0.88 (17)
Haryana	0.24 (28)	0.64 (20)	0.11 (28)	0.32 (28)	0.25 (25)	1.25 (9)	1.02 (19)	0.87 (19)
Himachal Pradesh	1.03 (12)	0.48 (27)	0.54 (12)	0.67 (15)	1 (12)	1.21 (11)	1.2 (9)	1.15 (9)
Jammu & Kashmir	0.06 (30)	1.09 (6)	0.07 (30)	0.4 (25)	0.05 (30)	1.22 (10)	0.86 (27)	0.74 (22)
Jharkhand	0.41 (24)	0.94 (10)	0.52 (14)	0.63 (17)	0.41 (21)	0.52 (28)	0.83 (28)	0.6 (29)

Karnataka	1.72 (7)	0.97 (9)	1.92 (4)	1.55 (6)	1.81 (7)	1.72 (6)	1.06 (17)	1.5 (6)
Kerala	3.03 (1)	0.65 (19)	1.39 (7)	1.64 (5)	2.85 (3)	2.04 (3)	1.46 (1)	2.06 (3)
Madhya Pradesh	0.3 (27)	0.4 (29)	0.24 (24)	0.31 (29)	0.24 (26)	0.59 (25)	0.9 (24)	0.61 (28)
Maharashtra	1.13 (11)	0.61 (21)	0.36 (19)	0.67 (14)	0.96 (13)	1.1 (12)	1.18 (10)	1.09 (12)
Manipur	0.72 (17)	0.74 (16)	0.32 (20)	0.58 (19)	0.69 (15)	0.28 (31)	1.12 (13)	0.72 (24)
Meghalaya	0.63 (19)	0.37 (30)	0.16 (26)	0.37 (26)	0.51 (19)	0.77 (19)	1.05 (18)	0.8 (21)
Mizoram	0.57 (20)	0.82 (14)	0.55 (11)	0.65 (16)	0.23 (27)	0.78 (18)	1.33 (2)	0.82 (20)
Nagaland	0.68 (18)	0.59 (22)	0.26 (23)	0.49 (22)	0.36 (24)	0.29 (30)	0.75 (31)	0.48 (31)
New Delhi	0.03 (31)	1 (8)	0.01 (31)	0.34 (27)	0.04 (31)	2.95 (2)	1.25 (7)	1.45 (7)
Odisha	1.81 (6)	0.67 (18)	1.5 (5)	1.32 (7)	1.84 (6)	0.56 (26)	1 (21)	1.1 (10)
Puducherry	2.93 (2)	1.39 (4)	2 (3)	2.08 (3)	3.11 (1)	2.03 (4)	1.29 (4)	2.07 (2)
Punjab	0.2 (29)	0.44 (28)	0.09 (29)	0.24 (31)	0.21 (28)	1.55 (7)	1.12 (12)	0.99 (14)
Rajasthan	0.53 (21)	0.55 (25)	0.31 (21)	0.45 (23)	0.6 (16)	0.69 (22)	0.79 (29)	0.7 (25)
Sikkim	1.19 (9)	0.55 (24)	0.37 (18)	0.67 (13)	1.25 (8)	0.83 (16)	1.2 (8)	1.09 (11)

Tamil Nadu	2.52 (4)	1.29 (5)	1.43 (6)	1.71 (4)	2.4 (4)	1.72 (5)	1.18 (11)	1.72 (5)
Tripura	1.34 (8)	2.85 (1)	4.64 (2)	3.07 (2)	1.08 (10)	0.81 (17)	1.28 (5)	1.07 (13)
Uttar Pradesh	0.33 (26)	0.54 (26)	0.4 (17)	0.42 (24)	0.2 (29)	0.75 (20)	0.89 (26)	0.64 (27)
Uttarakhand	0.48 (23)	0.76 (15)	0.52 (13)	0.59 (18)	0.4 (22)	1.05 (14)	1.09 (16)	0.87 (18)
West Bengal	1.03 (13)	1 (7)	0.57 (10)	0.85 (11)	1.12 (9)	0.67 (23)	1.11 (14)	0.97 (15)

Source: Author's calculation based on Status of Microfinance in India (NABARD, 2010. 2011), Basic Statistical Return (RBI, 2011), C.S.O, 2011 and Census, 2011.

Note: Figures in parenthesis indicates the corresponding rank of the state.

Notations: MP_1: Indicator of Microfinance Penetration, MP_2: Indicator of Microfinance Availability, MP_3: Indicator of Microfinance Usage, IMO: Index of Microfinance Outreach, WE_1: Economic Empowerment, WE_2: Financial Empowerment, WE_3: Social Empowerment, IWE: Index on Women Empowerment.

the penetration and availability of microfinance services actually determines their demand for microfinance. It is observed that the inequality in the usage of financial services is widespread across states of India. Once again, southen states (Andhra Pradesh, Puducherry and Karnataka) retains their supremacy in utilisation of microfinance services. Two north-eastern states, viz. Tripura and Odisha, perform better in utilising microfinance services. A poor rating of utilisation of microfinance products are observed in central, northrn, eastern and north-eastern states. Thus demand for microfinance services in these states are adequately addressed by the supply of those products. The combined measure of penetration, availability and usage of microfinance outreach is measured by an index of microfinance outreach. In terms of IMO, it is evident that only three states (Andhra Pradesh, Tripura and Pondicherry) have been classified under the category of high level of microfinance outreach. Five states (Tamil Nadu, Kerala, Karnataka, Odisha and Andaman & Nicobar Islands) have fulfilled the criteria of medium level of microfinance outreach as the value of IMO lying in between 1 and 2. All the other states belong to the lower stratum in the ladder of microfinance outreach.

Individual indicator wise analysis of women empowerment suggest that southern states like Andhra Pradesh, Karnataka, Kerala, Puducherry and Tamil Nadu perform relatively better in the ranking of economic and financial empowerment of women. Another important dimension of women empowerment in the social sphere is to expand women's access to literacy. Most of the union territories (such as Goa, Puducherry and Andaman & Nicobar) secure relatively higher ranking in the social empowerment. Overall, the predominance of southern states of India is established in the ranking of women empowerment index. In this regard, Goa occupied the highest ranking in the women empowerment index, followed by Puducherry and Kerala. Another ten states, viz. Andhra Pradesh, Tamil Nadu, Karnataka, New Delhi, Andaman & Nicobar, Himachal Pradesh, Odisha, Sikkim, Maharashtra and Tripura form the group of medium level of women empowerment with a value of women empowerment index ranging from 1.73 to 1.07. All the other 18 states represent low level of women empowerment holding index value lying within the range of 0.99 to 0.48.

5. Association between microfinance outreach and women empowerment: Indian experience

In order to examine whether there exist high degree of statistical correspondence between two sets of rank between IMO and IWE, the study estimated these two indices for 31 States and Union Territories in 2011-12. For the sake of analyzing the changes of the level of financial outreach and women empowerment over the period 2007-08 and 2011-12, the similar measurement is also carried out for 31 States and Union Territories in 2007-08.

Table 5.5 establishes a significant association between the estimated values of IMO and IWE. The estimated value of IMO varies between 11.66 in case of Tamil Nadu and 0.04 in case of Haryana in the year 2007-08. However, in the year 2011-12, IMO varies between 3.17 in case of Andhra Pradesh and 0.24 in case of Punjab. Among the better off States in respect of the level of microfinance outreach, Tamil Nadu and Puducherry had an IMO value of above 2 in 2007-08. States, like Andhra Pradesh, Tripura, Kerala, Karnataka, Odissa and Andaman &Nicobar Island were included in the category of high and medium level of microfinance outreach in the year 2011-12. A comparison of IMO with IWE suggests that IMO and IWE seem to move in the same direction. In the table, it has been pointed out that the ranks of IMO and IWE values for these 31 States and Union Territories move closely with each other[1]. The state, Puducherry, secure a high level of microfinance outreach and high women empowerment in both the state wise classification of 2007-08 and 2011-12. States having a medium level of microfinance outreach and a medium level of women empowerment are Tamil Nadu, Karnataka, Odisha and Andaman & Nicobar Island in the year 2011-12. However, a large number of states share the same characteristics of both the low level of microfinance outreach and low level of women empowerment in both the years 2007-08 and 2011-12 (see Appendix Table A1).

[1] The movement of both IMO and IWE values in 2011-12 is shown in Appendix figure A1. An association between these two indices is quite evident in the figure.

Table 5.5. Indices of microfinance outreach and women empowerment in 2007-08 and 2011-12

State	*2007-08*				*2011-12*			
	IMO Value	*Rank*	*IWE Value*	*Rank*	*IMO Value*	*Rank*	*IWE Value*	*Rank*
Andaman & Nicobar Islands	0.76	3	0.98	11	1.12	8	1.40	8
Andhra Pradesh	0.66	4	1.06	8	3.17	1	1.73	4
Arunachal Pradesh	0.22	15	0.59	27	0.31	30	0.73	23
Assam	0.10	22	0.65	23	0.88	10	0.89	16
Bihar	0.09	25	0.46	30	0.70	12	0.54	30
Chhattisgarh	0.23	12	0.60	25	0.56	20	0.68	26
Goa	0.27	8	2.70	3	0.93	9	2.49	1
Gujarat	0.08	27	0.84	15	0.52	21	0.88	17
Haryana	0.04	31	0.85	14	0.32	28	0.87	19
Himachal Pradesh	0.15	18	0.99	9	0.67	15	1.15	9
Jammu & Kashmir	0.06	29	0.78	19	0.40	25	0.74	22
Jharkhand	0.14	20	0.54	29	0.63	17	0.60	29
Karnataka	0.43	5	1.17	6	1.55	6	1.50	6
Kerala	0.35	6	1.43	5	1.64	5	2.06	3
Madhya Pradesh	0.09	24	0.57	28	0.31	29	0.61	28
Maharashtra	0.14	19	0.92	12	0.67	14	1.09	12
Manipur	0.16	17	0.60	24	0.58	19	0.72	24
Meghalaya	0.26	9	0.73	20	0.37	26	0.80	21
Mizoram	0.23	13	0.80	17	0.65	16	0.82	20

Nagaland	0.09	26	0.38	31	0.49	22	0.48	31
New Delhi	0.10	23	1.66	4	0.34	27	1.45	7
Odisha	0.32	7	0.70	21	1.32	7	1.10	10
Puducherry	6.17	2	19.08	1	2.08	3	2.07	2
Punjab	0.06	30	1.12	7	0.24	31	0.99	14
Rajasthan	0.11	21	0.60	26	0.45	23	0.70	25
Sikkim	0.21	16	0.80	18	0.67	13	1.09	11
Tamil Nadu	11.66	1	3.59	2	1.71	4	1.72	5
Tripura	0.24	10	0.88	13	3.07	2	1.07	13
Uttar Pradesh	0.06	28	0.66	22	0.42	24	0.64	27
Uttarakhand	0.22	14	0.81	16	0.59	18	0.87	18
West Bengal	0.23	11	0.98	10	0.85	11	0.97	15

Source: Authors calculation based on Basic Statistical Return of SCBs (Reserve Bank of India) 1993 and 2004, Economic Survey, 2004-05 and 2009-10. The values of HDI components were collected from NSSO 50th and 61st Round EUS; NFHS I & III; RGI, 2006.

The outreach of microfinance programme is likely to expand better access to economic opportunities among SHG members. As the programme is often targeted to an overall development of women sections of the society, so it is expected that the programme would address the empowerment of women. In an effort to integrate these issues, Figure 5.1 simply plot the relationship between the outreach of microfinance services and the level of women empowerment across states of India in 2011-12. In this figure, the scatter dots represent the observations of various states. Given the two-way causality relationship between microfinance outreach and women empowerment, the north-east quadrant represents the virtuous cycle, i.e., as we move into the ranges of states with very high level of microfinance outreach, the process of women empowerment will be at a high levels as well. In the south-west quadrant, states having low microfinance outreach may result in low women empowerment, i.e., a situation of vicious cycle. The south-east and north-west quadrant represent the lopsided-microfinance outreach and lopsided-women empowerment respectively. In these situations, good performance in one dimension may not ensure the better performance in the other. The majority of observations lie within the southwest portion, i.e., in the category of vicious cycle of the scatter diagram. It implies that lower outreach of microfinance programme is significantly correlated with lower level of women empowerment. Interestingly, five southern states (Andhra Pradesh, Puducherry, Tamil Nadu, Kerala and Karnataka) experienced a robust outreach in microfinance programme, which eventually assisted the movement towards virtuous-cycle category. Outreach of microfinance programme also helps the state, Tripura, to move marginally to lopsided microfinance outreach category, even though the performance of the state is not satisfactory in achieving empowerment of women. On the other hand, Goa, New Delhi and Andaman & Nicobar Island secure a higher ranking in women empowerment. Widening of financial and social empowerment without any significant outreach of microfinance programme, enable these states to move to the lopsided women empowerment category. Overall, it can be suggested that the level of women empowerment is a powerful correlate of microfinance outreach. The evidence is expected to found enough empirical support if we rank states rather than cardinal measures[1]. In other words, if we rank states according to their level of microfinance outreach and then compute similar

[1] Dasgupta (1993) showed that per capita income is correlated even more highly with other indicators of development if we consider ranks rather than cardinal measures.

ranks based on women empowerment, then we find a high degree of statistical correspondence between the two sets of ranks. In fact, in our study, some statistical evidences also suggest that the values of IMO and IWE for the set of states move closely with each other. The estimated value of correlation coefficients between the components of IMO and the IWE in 2011-12 is shown in the Appendix table A2. It can be seen that the correlation coefficients between IMO and IWE are estimated to be about .564 (Pearson Correlation), .445 (Kendall's tau-b) and .634 (Spearman's rho) in the year 2011-12. All these coefficients are found highly statistically significant at 1 percent level of significance (2-tailed). Moreover, it is also evident that most of the components of IMO and IWE are highly correlated among each other. Overall, it can be concluded that states having high level of microfinance outreach are also the states with a relatively high level of women empowerment.

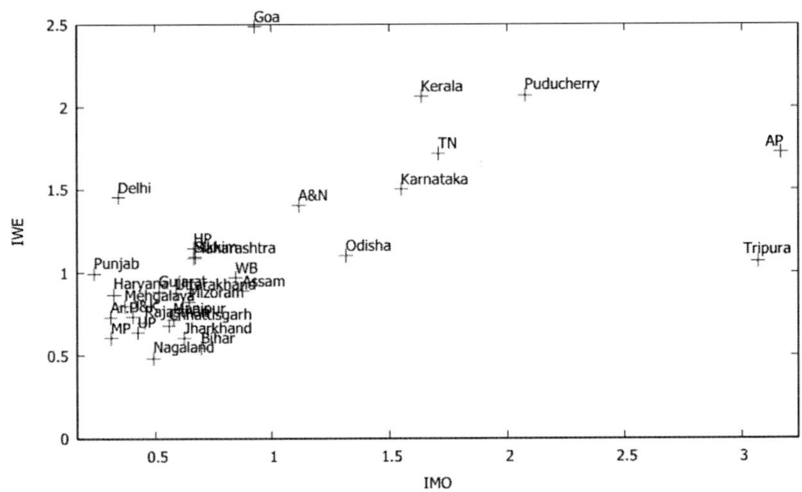

Figure 5.1. Scatter Plot of IMO and IWE in 2011-12

6. Conclusions and policy implications

The paper focuses the linkage between microfinance outreach and women empowerment. In an effort to establish the link, two comprehensive measures- index of microfinance outreach and women empowerment index have been constructed. An analysis on the inter-state variation in the microfinance outreach suggest that only in eight states (Andhra Pradesh, Tripura, Puducherry, Tamil Nadu, Kerala, Karnataka, Odisha and Andaman & Nicobar Island) outreach of microfinance programme is significant. Interestingly, among these eight states, Puducherry Tamil Nadu, Karnataka, Odisha and Andaman & Nicobar Island have witnessed a higher and medium level of women's empowerment. Thus, women empowerment is found to be a powerful correlate of microfinance outreach among the poor households. However, there is wide variability in the outreach of microfinance across various regions of India. The southern region is leading in the outreach of microfinance programme, followed by central, northern, north-eastern and eastern regions. It is thus desirable to create conditions for enhancing the outreach of microfinance programme especially to low inclusive states so as to reduce the regional imbalances. It is predicted that an all-inclusive microfinance system would strengthen the process of financial inclusion in India and thereby would promote women's empowerment.

Appendix

Table A1. Correlation matrix of different components of IMO and IWE, 2007–08

		MP_1	MP_2	MP_3	IMO
Pearson correlation	WE_1	.660** (.000)	.037 (.843)	.990** (.000)	.994** (.000)
	WE_2	.141 (.449)	.085 (.650)	.143 (.444)	.145 (.435)
	WE_3	.233 (.208)	.231 (.211)	.215 (.245)	.220 (.234)
	IWE	.661** (.000)	.051 (.785)	.981** (.000)	.985** (.000)
Kendall's tau_b	WE_1	.901** (.000)	.185 (.144)	.557** (.000)	.768** (.000)
	WE_2	.080 (.529)	.181 (.153)	.140 (.269)	.161 (.202)
	WE_3	.286* (.024)	.129 (.308)	.286* (.024)	.333** (.008)
	IWE	.286* (.024)	.181 (.153)	.286* (.024)	.342** (.007)
Spearman's rho	WE_1	.980** (.000)	.239 (.196)	.722** (.000)	.905** (.000)
	WE_2	.151 (.417)	.281 (.125)	.198 (.286)	.244 (.187)
	WE_3	.406* (.024)	.221 (.233)	.404* (.024)	.481** (.006)
	IWE	.411* (.022)	.288 (.116)	.394* (.028)	.479** (.006)

Note: Figures in parenthesis indicate significant level in two-tailed test

* Correlation is significant at the 0.05 level (2-tailed)

** Correlation is significant at the 0.01 level (2-tailed)

IP: Index for Penetration; IA: Index for Availability; IU: Index for Usage; MPCE: Monthly Per Capita Expenditure; HEALTH: Indicator of Health Attainment; EDU: Indicator on Educational Attainment.

Table A2. Correlation matrix of different components of IMO and IWE, 2011–12

		MP_1	MP_2	MP_3	IMO
Pearson correlation	WE_1	.984** (.000)	.337 (.064)	.635** (.000)	.784** (.000)
	WE_2	.187 (.314)	.366* (.043)	.036 (.846)	.166 (.371)
	WE_3	.468** (.008)	.330 (.070)	.166 (.371)	.330 (.070)
	IWE	.702** (.000)	.456* (.010)	.381* (.035)	.564** (.001)
Kendall's tau_b	WE_1	.862** (.000)	.209 (.099)	.553** (.000)	.656** (.000)
	WE_2	.148 (.241)	.243 (.055)	-.015 (.905)	.140 (.269)
	WE_3	.290* (.022)	.222 (.080)	.144 (.255)	.273* (.031)
	IWE	.488** (.000)	.333** (.008)	.256* (.043)	.445** (.000)
Spearman's rho	WE_1	.961** (.000)	.317 (.083)	.727** (.000)	.841** (.000)
	WE_2	.217 (.242)	.362* (.045)	.016 (.931)	.240 (.194)
	WE_3	.434* (.015)	.307 (.093)	.217 (.242)	.422* (.018)
	IWE	.651** (.000)	.463** (.009)	.396* (.028)	.634** (.000)

Note: Figures in parenthesis indicate significant level in two-tailed test
* Correlation is significant at the 0.05 level (2-tailed)
** Correlation is significant at the 0.01 level (2-tailed)
IP: Index for Penetration; IA: Index for Availability; IU: Index for Usage; MPCE: Monthly Per Capita Expenditure; HEALTH: Indicator of Health Attainment; EDU: Indicator on Educational Attainment.

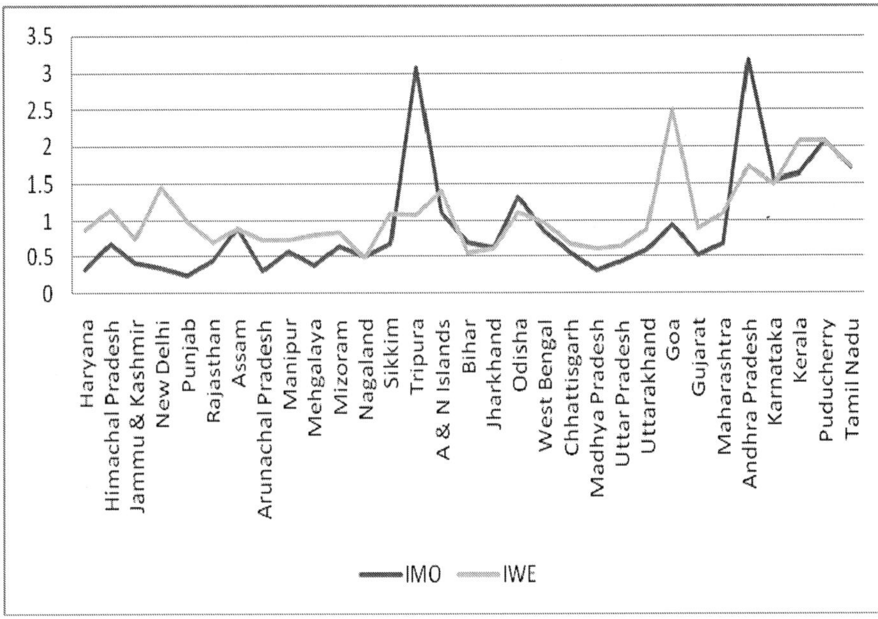

Figure A1. Movement of IMO and IWE in 2011-12

References

Burra, N., R.K. Murthy & J.D. Ranadive. *Micro Credit, Poverty and Empowerment*, Sage Publication, 2005.

Dasgupta, P. *An enquiry into Well-Being and Destitution*, Oxford: Clarendon Press, 1993.

Government of India. *Economic Survey, 2011-12.*

Hirway, I & D Mahadevia. "Critique of Gender Development Index-Towards an Alternative", *Economic and Political Weekly*, October 26, 1996.

Maes, J.P & L.R. Reed. *State of the Microcredit Summit Campaign Report 2012,* Published by the Microcredit Summit Campaign, Washington, United States of America, 2012.

Malhotra, A., S. R. Schuler & C. Boender. *Measuring Women's Empowerment as a Variable in International Development,* The World Bank, 2002.

Mehta, A.K. "Recasting Indices for Developing Countries-A Gender Empowerment Measure", *Economic and Political Weekly*, October 26, 1996.

NABARD. *Report of the Committee on Financial Inclusion*, January 2008.

—. *Status of Microfinance in India, 2011-12,* NABARD, 2012.

Pitt, M.M & S.R. Khandker. "The Impact of Group Based Credit Programs on Poor Households in Bangladesh: Does the Gender of Participants Matter?", *Journal of Political Economy*, 106 (5), 1998.

Sangwan, S.S. *Financial Inclusion and Self Help Groups*, 2008 www.nabard.org/.../Financial%20lnclusion%20and%20SHGs.pdf

Srinivasan, N. *Microfinance India-State of the Sector Report*, Sage Publications India Pvt. Ltd., 2008.

—. *Microfinance India-State of the Sector Report*, Sage Publications India Pvt. Ltd., 2009.

Verman, M.P. "Impact of Self Help Groups on Formal Banking Habits", *Economic and Political Weekly*, April 23, 1705-13, 2005.

World Bank. *Empowerment and Poverty Reduction: A Sourcebook*, 2002

Zubair, M. Women Empowerment through Micro-credit, in Gandhi edited *Women's Work, Health and Empowerment*, Aakar Books Publication, Delhi, 2006.

CHAPTER SIX

GENDER DIFFERENTIAL IN SOCIO-ECONOMIC CAPABILITIES AND ISSUES OF LIVELIHOOD DIVERSIFICATION

SOUMYENDRA KISHORE DATTA AND TANUSHREE DE

1. Introduction

In rural areas of developing economies, women constitute the driving force of development in their communities. They often combine the role of farm labourer or an off-farm occupation with household labour, and are often entrepreneurial cash-earners supporting their families and creating opportunities for others. They have to perform activities like cooking, caring for children and the elderly, collecting water and wood and the overall management of the household. It is the rural women who, from the early morning, have to take the most constrained decision about how to provide food for their children and adult members throughout the day with very limited means, how to maintain a balance between daily family consumption and the resources at hand, how to allocate time between diverse activities within and outside the home, and how to maintain order and peace in the family and societal ambience. There is no doubt that because of their greater presence within the household compared to men, women are the household managers from several angles. Hence, it is imperative that they be empowered physically, mentally and economically to act as potential change agents in the rural sphere by motivating their children and spouses to act in the right direction. Yet in many instances in the developing regions it is found that they still have far less access to material resources, education and training, health related benefits and economic and employment opportunities. Particularly, they remain deprived in terms of provision of health and education services for women.

Several constraining factors operate that maintain their capabilities at a subdued level compared to males, and in the process females usually do not get opportunities to pursue a more diverse form of livelihood. This often stands in the way of their continuous earning and affects their financial independence and social dignity. Sometimes, the gender differential in socio-economic attainments is so acute that women face severe problems in earning from a diversified access to work opportunities in their social surroundings. In this context it seems imperative:

(i) To analyse the association between gender gap in capabilities and the extent of diversification in jobs.
(ii) To focus on the livelihood diversification strategies pursued by males and females and associated income earnings.
(iii) To analyse the effect of income diversification index, gender aspects and other factors that explain individual income earnings across the study region.
(iv) To analyse the differential status of ownership of resources like available land, labour and livestock etc., and subdued social mobility that caters to the discriminatory status of female livelihood diversification and their relatively lower empowerment.

2. Literature Review

Gender differences in various levels of achievement emerge from the socially structured relationship between men and women (Oakley 1972). Inequality can result from differences in either efforts, which are under the control of an individual, or circumstances such as gender, religious background, geographical location, and parental education, which are beyond the control of the individual (Roemer 2006).These differences affect the distribution of resources and responsibilities between men and women, and are shaped by ideological, religious, ethnic, economic and cultural determinants (Moser 1989). Being socially determined, this distribution can thus change through conscious social action, including public policy (Quisumbing 1996). There are a number of studies and evidences which document the gender gap/inequality at various levels of achievement, for instance in a World Bank study (2006) it is reported that gender inequality is the typical "inequality trap," which has been caused and reinforced by interlinked cultural, social and economic factors within and outside the household.

An FAO paper (2011) focuses on gender differential in asset ownership across several developing countries of the world, and it is observed that females and female-headed households and farms lag behind their male counterparts in their access to and ownership of most inputs, assets and services that shape the productive activities in rural areas. Although in terms of education female attainment is sometimes on apar with that of males, but in most regions even in this area, females lag behind. This skewed distribution of assets has an untoward impact on the socio-economic mobility of females and damages their social welfare.

3. The study area

Against this backdrop it seems imperative to analyse the issue of gender differential in capability and in diverse aspects of livelihood, as well as the diversification strategies followed by people in four villages in a backward district in West Bengal. A wide variety of data on many aspects (i.e. employment status, sectors of employment, earnings, educational level, health status, wealth level, demographic features, mobility status etc.) at the household as well as at individual levels were collected to understand the complexity of capability pattern and rural livelihoods and the gender issues governing the male-female divide in earning from a diversified livelihood base. The data are varied in dimension and there is no way to find this disaggregated form of data from secondary sources, except for undertaking primary survey. The data were collected on a purposive random sampling basis from four villages: Paharpur and Mana under the Borjora gram panchayat, and Katabundh and Kotalpukur in the Hat-Asuriya gram panchayat in the Bankura district of West Bengal. From each village 40 households were surveyed giving a total of 160 sample households, while corresponding to employed individual members the number is far greater.

4. Relation between gender gap in capabilities and diversification indices

It is important to note that people's capacity to resort to diversified occupational patterns is governed by their degree of access to relevant resources, educational and vocational skill, physical ability and socio-economic opportunities. It is said that human capability is conditioned by relative attainment with regard to three components: education level, health status and income status, that may be taken as a proxy of the socio economic conditions favouring their access to resources. The socio-cultural background in rural regions and lack of proper Govt. initiatives often tend to keep the attainment level of female capabilities in the aforesaid fields at a suppressed level compared to their male counterparts. In this context it seems pertinent to focus on possible gender gap in capability across the study region and relate it to the gender gap in diversification indices.

It needs to be mentioned in this context that the local government has often bypassed its commitments to making a provision of continuing education facilities for adult women through stressing the creation of SHG groups. Apart from this, little attention is given to the specific problems of school dropouts of female children without pondering the relevant causes in an analytical and systematic manner. There has been a glaring neglect of the diverse aspects of female education in the rural sphere, particularly in the perspective of globalization. Together with a very casual approach to the aspect of general education, no serious attention has been given to the skill development and imparting of training facilities to women inthe countryside commensurate with available local resources and employment prospects. The few programmes that have been launched pertaining to women and education are hardly sensitive to the needs of learners. As a result, women suffer in a system of education that thwarts their socio-economic mobility and implicates an inertia of low capability, low information base and low access to better employment prospects in and outside their local sphere. Apart from this, women are often victims of the poorer provision of health care facilities that are severely needed to sustain their physical strength, stamina and vigour for outdoor jobs, aside from their engagement in daily household chores. The compulsions of child bearing without an adequate nutritional diet and medical support, cooking with bio-fuels that lead to the inhalation of toxic gases, lack of respite from day-long family jobs, including care for elderly and children, and collection of fue land water etc. leave them exhausted and drained of inner strength and energy. The gender gap in the health care facility exacerbates this aspect of female deprivation. Besides this, socio-economic taboos and differential opportunities of employment tend to confine the women to lower income earning jobs. In order to capture the gender gap in capabilities, an index is constructed focusing on the relative differential in the attainment of women *vis-a-vis* men in terms of education, health and income aspects. For this purpose, males and females in each family are divided into two separate groups. Now, from each specific group the average years of schooling are calculated by summing the total years of schooling of the relevant family members in each group and then dividing by the number of included individuals. Thus, if in family i, X_i is the average years of schooling for any gender group, then considering all the families in the sample $\frac{(X_i - X_{min})}{(X_{max} - X_{min})}$ is taken as the index of educational attainment for the ith family in that gender group. A similar index is constructed by considering the average health care expenses as well as average income of any gender group in each family i in the sample. The overall average of these indices is taken as the index of attainment of capability in any particular gender in the ith family. The difference between the female capability index and male capability index in the ith family is considered as the corresponding index of gender gap in capability. Again, a diversification index for each gender group in the ith family is computed by three alternative methods:

(i) by taking the inverse of Hirschman–Herfindahl index Where $H - Hindex = \sum S_{ij}^2$, S_{ij} being the share of income of any particular gender in family i from the jth occupation

(ii) by applying Shannon's entropy formula where $Shannon's\_index = -\sum S_{ij} \ln(S_{ij})$, and

(iii) by applying Berry index in the form $1 - \sum S_{ij}^2$.

Females have a comparably lower earning potential than males because of their relatively lower physical strength for harder manual work, less socio-economic mobility due to socio-cultural taboos, and a built-in social bias against female employment except in occupations requiring low-paid female labour. Traditionally, women have suffered from a segmented labour market with employment prospects only in low-paid, agriculture-related work. However, with the spread of mechanization, women have been reduced to the status of low-skilled and low-wage labour compared to their male counterparts. Although in recent times both female and male employment scope has broadened with the advent of NREGS, together with 33% of jobs being reserved for women, there has not been a commensurate realized employment gain for the women. Women in relatively well-educated families and Muslim dominated regions are contained in their efforts to opt for an NREGS job due to pressure from other members of the household. The main obstacle that the women usually have to face is the socio-cultural stigma attached to physical work and working alongside male members in the community. This, along with their relatively low skills and training, usually continue to keep their income earnings at a depressed level. Table 6.1 below provides a picture of the sympathetic movement of gender gap indices for capability, and that of diversification indices measured in three different ways. In order to focus on gender gap in the diversification index, only earning males and females have been considered. Referring to the Borjora panchayat, the fifth column gives a synoptic view of the matching cases of negative values of the gender gap index in capability with negative values of diversification indices measured in three different ways, while the sixth column provides similar cases of matching in respect of positive values of both. The almost uniform value, in either negative or positive cases for the three indices, suggests the reliability of the data despite its being collected from door-to-door surveys. The values of gender gap indices in capability are observed to be somewhat moderately but significantly correlated with the gender gap indices in diversification, measured by all the three methods in cases of both panchayats. This is indicative of the fact that as women fail to catch males in terms of socioeconomic capability because of their relatively weaker education and health statuses, they also lag in respect of earning income from varied sources due to difficulty in access to productive resources, lack of proper skills, training and physical stamina, and vice versa. Again, as females in general lag behind the male counterparts in terms of their relatively weak participation in the non-farm job sphere, they suffer in terms of lack of access to the ready financial resources at hand and their weaker voices in the family decision making, which in turn relegate their socio-physical state.

Table 6.1. Correlation across gender gap in capability and gender gap in diversification indices

Gram Panchayat	Measure of Gender Gap in Diversification	Correlation with Gender Gap in Capability Index	Significant	Both Negative	Both Positive
Barjora	Shannon Index	0.350	0.001	30	6
	Inv HH Index	0.560	0.000	37	6
	Berry Index	0.368	0.001	30	6
Hat-Asuriya	Shannon Index	0.412	0.000	41	15
	Inv HH Index	0.591	0.000	44	15
	Berry Index	0.399	0.000	39	14

#Source: Field Survey 2012–13

5. Livelihood diversification strategies and income

Table 6.2 below displays the gendered percentage distribution of individuals according to dimensions of activity pursued in the two panchayats, and ably demonstrates the fact that adopting diverse activities has become an important source of earning for some people. For purposes of expositional convenience we categorise the individuals into three broad groups: (i) individuals that solely depend on farming and allied jobs; (ii) individuals that pursue both farming and non-farming jobs; (iii) individuals adopting only non-farming activities. The types of non-farming activities available in rural areas can largely be divided into five categories: self-employed enterprise, small business, production, casual labour and formal /informal employment (Smith et al. 2001; Ellis 2005). Accordingly, in the study area small business includes works like rice business, vegetable vending, grocery, saree business etc. Casual labour refers to khalasi, contract labour, driving, work in brick kiln, masonry etc.; production includes carpentry, fabric work, balaposh making etc.; deputed teacher, Anganwari workers, NREGA constitute formal employment while SHGs activity constitute self-employed work.

Table 6.2 shows that pure farm dependence is relatively low for males (18.71%) compared to females (37.68%) in the Barjora panchayat. If dependence on non-farm activities is considered together with the farming sector, females (62.32%) are observed to substantially dominate in this category. This demonstrates that females are mostly concentrated in the low-income, pure farming sector, since in the mixed case (farm + non-farm) male dependence is found to be somewhat higher (26.62%) compared to females (24.64%).In the case of only non-farm activities, however, male dependence is found to be far larger (54.68%) compared to females (37.68%). These figures are clearly indicative of major female dependence mostly on agriculture-related jobs, while males mostly seek to diversify into high return non-farm areas. In the case of the Hat-Asuriya Panchayat, however, pure farm dependence for females is observed to be nil while for males it is just 9.09%. In the case of mixed dependence, combining the farm and non-farm sectors together, female dependence is observed to be somewhat larger (76.67%) in comparison to the males (63.64%). In the case of the non-farm sector alone, however, male diversification is dominant (24.81%) over females (16.98%). If pure non-farm dependence is considered together with the aforesaid combined case, then the figure for the females stands at 100%, while for males it is 92.25%.

Table 6.2. Percentage dependence of females and males on farm and non-farm sectors

Activity Status	Barjora Gram Panchayat		Hat-Asuriya Gram Panchayat	
	Female	Male	Female	Male
Farm Only	26 (37.68%)	26 (18.71%)	N.A.	12 (9.09%)
Farm and One More Occupation	17 (24.64%)	32 (23.02%)	20 (33.33%)	22 (16.67%)
Farm and Two More Occupations	N.A.	5 (3.6%)	23 (38.33%)	58 (43.94%)
Farm and Three More Occupations	N.A.	N.A.	3 (5.00%)	4 (3.03%)
Without Farm Only One Occupation	23 (33.33%)	70 (50.36%)	12 (20.00%)	20 (15.15%)
Without Farm Two More Occupations	3 (4.35%)	6 (4.32%)	2 (3.33%)	16 (12.12%)
Total Employed People	69 (100.00%)	139 (100.00%)	60 (100.00%)	132 (100.00%)

Source: Field Survey 2012–13

It also seems of interest to analyse the livelihood strategies for both genders to find the relative importance of farm income *vis-a-vis* non-farm income across different income groups.

Table 6.3. Gender-specific income share from farm and non-farm sectors across income groups (figures in percentages)

Gram Panchayat	Village	Economic Sector	Gender	less than Rs. 10,000	Rs. 10,001 Rs. 25,000	Rs. 25,001 Rs. 50,000	Rs. 50,000 Rs. 75,000	Rs. 75,001 Rs. 100,000	Above Rs. 100,000
Barjora	Paharpur	Farm	F	0.00	56.23	21.26	18.03	0.00	N.A
			M	100.00	43.77	78.74	81.97	100.00	N.A
		Non-farm	F	98.07	34.01	25.20	34.35	0.00	0.00
			M	1.93	65.99	74.80	65.65	100.00	100.00
	Mana	Farm	F	N.A	62.07	35.34	0.00	0.00	0.00
			M	N.A	37.93	64.66	100.00	100.00	100.00
		Non-farm	F	76.79	41.90	10.20	0.00	0.00	0.00
			M	23.21	58.10	89.80	100.00	100.00	100.00
Hat-Asuriya	Katabandh	Farm	F	29.23	41.67	0.00	0.00	N.A	N.A
			M	70.77	58.33	100.00	100.00	N.A	N.A
		Non-farm	F	51.61	42.86	4.35	0.00	0.00	0.00
			M	48.39	57.14	95.65	100.00	100.00	100.00
	Kotalpukur	Farm	F	32.14	28.57	0.00	0.00	N.A	0.00
			M	67.86	71.43	100.00	100.00	N.A	100.00
		Non-farm	F	42.86	45.45	13.51	0.00	0.00	N.A
			M	57.14	54.55	86.49	100.00	100.00	N.A

Source: Field Survey 2012–13

It is interesting to note from Table 6.3 above that in the lower income groups across the two villages in the Barjora panchayat, the share of women regarding farm income is greater than that of men. Even at the lowest income section, the female share of income from low productive non-farm activities exceeds that of males as revealed in the case of the Mana village in the Barjora Panchayat and the Katabundh village in the Hat-Asuriya panchayat. This is explained by the fact that a greater part of the sampled households are small land holders or marginal farmers. Since the productivity of small holdings is usually low and capital intensive cultivation is hardly possible here, the males usually diversify to relatively high income yielding non-farm jobs while females are pushed to the low productive farm operations on a bigger scale. In case of both the villages in the Hat-Asuriya panchayat, however, the male share of farm income even at lower income groups dominate over females. However, with the rising income groups, the female share in both farm and non-farm sectors begins to decline in every village while that of males in both the cases increases. Even in some of the cases in upper income groups, female activity is non-existent. This is a clear pointer to the fact that female participation in the labour market remains segmented and its presence seems to be crowded mostly in low-return farm and non-farming jobs.

Despite the fact that for poorer women the pursuit of varied activities helps supplement household income in cases of dire necessities, the participation of women in spheres outside the traditional farm sector is often neither automatically guaranteed on a continuous basis, nor are they mentally and physically adapted to taking the challenge of multiple types of jobs that often require a special type of skills and training. Apart from this, the continuity of female labour in the non-farming sector is often conditioned by the physical circumstances and patriarchal forces that operate both at the workplace and in the particular household they belong to (Jeyaranjan & Swaminathan 1999).

The survey reveals that females in the villages are mostly engaged in low-skill activities like agricultural labour, NREGA, daily wage labour, cooking (SHG activity) or balaposh making. In contrast, the males in both cases are employed in a diverse array of activities like fishing, NREGA, masonry, groceries, driving, agricultural labour, business, teaching etc., which require certain amounts of skill and yield better returns than the jobs females are usually engaged in. Apart from this, females often remain deprived of the ownership of bigger plots of land and access to diverse forms of assets, which stifle their better work participation and higher income earning capacities by following a diversified mode of livelihood.

6. Impact of Diversification, Gender and Other Issues on Individual Income

The primary condition for earning-increased income lays in having access to an enhanced asset base and opportunities for pursuing diverse livelihood options. An individual who can enjoy increased command or ownership of diverse types of assets (e.g. building a more secure house, increasing livestock herds, owning agricultural land or having increased access to information through social networks) is likely to have more opportunities to diversify their activities and earn a more steady and secured income than a person with poor access to assets and a low diversified portfolio of income. Thus, livelihood activity options are dependent on an individual's assets and their ability to convert assets to activities (Ellis 2000; Rakodi 2002).In this section the impact of individual diversification, and other socio-economic features on individual income are analysed. The regression is of the following form:

$$IND\_INCOME_i = \alpha_0 + \alpha_1(MOD\_CST)_i + \alpha_2(AGE)_i + \alpha_3(SQAGE)_i + \alpha_4(GEND)_i + \alpha_5(EDU)_i + \alpha_6(DIV\_IND)$$

Here, the DIV-IND index indicates the individual diversification index. Three types of diversification indices—Shannon index, inv-HH (inverse of Hirschman-Herfindahl) index and Berry index—are considered in the regression to reflect the impact of diversification (measured in different ways) on income. It is assumed that as diversification rises, the income-earning prospect also rises through access to varied sources. The education of individuals, denoted by EDU and measured by years of schooling, is supposed to have a positive relation to income. Gender is a qualitative variable assuming value 1 if the individual is female and 0 if male. In the patriarchal society, socio-cultural taboos imposed on women are supposed to constrain their diversification opportunities and hence lower income prospects relative to their male counterparts. So, males are assumed to earn more than females, on average. Age is assumed to have a U-shaped impact on earnings. Modified general caste people are indicated by 1 while non-general caste people are indicated by 0. Income earning is likely to be higher for the general category because of their superior education, connection and social networks.

Table 6.4. Regression results of individual incomes

Explanatory Variables	Coefficient	Explanatory Variables	Coefficient	Explanatory Variables	Coefficient
Modified_Caste	1584.25	Modified_Caste	1638.274	Modified_Caste	1626.732
Age	1034.337*	Age	1045.396*	Age	1044.697*
Squ_Age	(-9.499)**	Squ_Age	(-9.664)*	Squ_Age	(-9.632)*
Gender	-18857.66*	Gender	-19149.25*	Gender	-18900.23*
edu_quali	1423.038*	edu_quali	1371.294*	edu_quali	1408.101*
Shannon_Index	7308.737**	Inv -HH_Index	4394.289*	Berry_Index	11140.8*
Constant	-8154.038	Constant	-12002.801	Constant	-8170.675
R2	0.233	R2	0.229	R2	0.232
F	19.563	F	19.21	F	19.449

Source: Field Survey 2012–13
General= Hindu General+ Muslim general; Non-General= Hindu, Muslim Non General

The value of the F statistic indicates that all the regressions are good fits. Again, all three types of diversification indices are observed to have significant positive impacts on levels of income earnings, implying that as diversified sources of employment rise, income also goes up. Age has a significant positive impact on earnings in the case of all three regressions, while the rate of increase falls with increase in age as evident from the significant negative value of the SQU AGE variable. Education (EDU), as expected, is found to have a positive and significant impact on income. Gender reflects the expected negative sign in all cases. This is indicative of the fact that women are confronted with relatively more constraints in diversified income earning avenues than men. Modified caste has the expected positive value in all the cases; however, these are not significant.

7. Asset profile and gender bias

Ownership of/access to key-assets (such as savings, homestead, land, livestock, jewellery, education, training, and access to market and employment information, right to use common property natural resources etc.) is a an essential requirement that enables rural households and individuals to diversify in various activities (Dercon & Krishan 1996; Abdulai & Crole Rees 2001). Investment of a proper mix of the above endowments is the starting move of any independent activity.

Bebbington (1999) provides a broad understanding of assets. According to him: "Assets are not simply resources that people use in building livelihoods: they are assets that give them the capability to be and to act. Assets should not be understood only as things that allow survival, adaptation and poverty eradication: they are also the basis of agents' power to act and to reproduce, challenge or change the rules that govern the control, use and transformation of resources." The term "access" indicates the social dynamics of institutions that govern the relative control exercised by different groups of people over available resources. Ian Scoones (1998) defines access as "the rules and social norms that determine the differential ability of people in rural areas to own, control, otherwise claim or make use of resources such as land and common property." Not everybody in the society or even within the household enjoys equal access to assets. An individual's access is determined by the social institutions or relations, or their social and demographic status within households, comprising factors like gender, caste, class, age, religion etc.

The lower level of earnings and relatively subdued form of employment diversification on the part of females compared to males can partly be attributed to the gender difference in access to and ownership of diverse tangible assets, and intangibles like information and education in the study areas. At present, the state of having access to information and increased mobility greatly influence the pattern of adoption and switching across diverse forms of jobs. In this context, the focus is on education, residence, agricultural land, livestock, assets regulating access to information and mobility, as well as assets like jewellery that provide some sort of emotional fervour and security in the mindsets of women. Table 6.5 below focuses on the gender specific relative ownership of diverse types of assets in the four villages of the two surveyed panchayats.

In Paharpur only 22.50% of women out of 40 households, and in Mana only 5% women out of 40 households, enjoy ownership of the houses they reside in. The corresponding percentages are just 10% and 2.5% in Katabundh and Kotalpukur, respectively, in the Hat-Asuriya panchayat. The deprivation of female members is evident in the ownership of agricultural land, as only 14.29 % and 4.76 % of women have such ownership rights in Paharpur and Mana, while the corresponding figures are only 10% and 5% in Katabundh and Kotalpukur. The figures are much lower compared to male ownership of such land. The laws of inheritance are against the female family members, who cannot raise an effective clamour to establish their rights over land. This bleak land ownership status of women is commensurate with the findings of Rao (2013), who in a study in Jharkhand found that women claiming a share of land were subdued by socially demeaning brands. Despite the enunciation of the Hindu Succession Act in 2005, through which daughters were granted an equal share as sons, hardly any tangible change is in sight. The social perception among men and women vindicates male inheritance and the maleness of land as an asset. In all the villages women SHGs exist to some extent and mostly undertake livestock rearing and midday meal cooking activities. In terms of ownership of livestock, females and males have an almost similar standing in the Borjora panchayat. In the case of villages in the Hat-Asuriya panchayat this ownership of cattle is distributed in favour of the males. In terms of percentages of possessing cell phone or vehicles, females lag far behind the males in the villages as the data in Table 6.5 reveal. This is supposed to seriously impair their mobility and act as constraints in undertaking a diversified livelihood option, even if they may have the desire to do so. Only in terms of possessing jewellery do females far outnumber males. However, since this hardly helps in generating any income, it does not assist in adopting any diversified livelihood and acts only as some sort of security during extremely hard times. Again, while ownership of a diverse array of asset bases may impact on the earning

Table 6.5. Distribution of assets across gender in the surveyed villages in the Borjora Block

Ownership	Gram Panchayat	Village	No Asset Holder	Asset Holder	Men	Women	Both Men & Women
Residential	Barjora	Paharpur	0 (0.00%)	40 (100.00%)	29 (72.50%)	9 (22.50%)	2 (5.00%)
		Mana	0 (0.00%)	40 (100.00%)	37 (92.50%)	2 (5.00%)	1 (2.50%)
	Hat-Asuriya	Katabandh	0 (0.00%)	40 (100.00%)	32 (80.00%)	4 (10.00%)	4 (10.00%)
		Kotalpukur	0 (0.00%)	40 (100.00%)	36 (90.00%)	1 (2.50%)	3 (7.50%)
Agriculture Land	Barjora	Paharpur	33 (82.50%)	7 (17.50%)	6 (85.71%)	1 (14.29%)	N.A
		Mana	19 (47.50%)	21 (52.50%)	19 (90.48%)	1 (4.76%)	1 (4.76%)
	Hat-Asuriya	Katabandh	20 (50.00%)	20 (50.00%)	15 (75.00%)	2 (10.00%)	3 (15.00%)
		Kotalpukur	20 (50.00%)	20 (50.00%)	18 (90.00%)	1 (5.00%)	1 (5.00%)
Livestock	Barjora	Paharpur	25 (62.50%)	15 (37.50%)	8 (53.33%)	7 (46.67%)	N.A.
		Mana	13 (32.50%)	27 (67.50%)	13 (48.15%)	12 (44.44%)	2 (7.41%)
	Hat-Asuriya	Katabandh	26 (65.00%)	14 (35.00%)	9 (64.29%)	4 (28.57%)	1 (7.14%)
		Kotalpukur	22 (55.00%)	18 (45.00%)	13 (72.22%)	3 (16.67%)	2 (11.11%)

Jewellery	**Barjora**	Paharpur	27 (67.50%)	13 (32.50%)	1 (7.69%)	12 (92.31%)	N.A.
		Mana	32 (80.00%)	8 (20.00%)	N.A.	7 (87.50%)	1 (12.50%)
	Hat-Asuriya	Katabandh	26 (65.00%)	14 (35.00%)	2 (14.29%)	12 (85.71%)	N.A
		Kotalpukur	24 (60.00%)	16 (40.00%)	2 (12.50%)	13 (81.25%)	1 (6.25%)
Vehicle	**Barjora**	Paharpur	30 (75.00%)	10 (25.00%)	9 (90.00%)	N.A.	1 (10.00%)
		Mana	17 (42.50%)	23 (57.50%)	16 (69.57%)	1 (4.35%)	6 (26.09%)
	Hat-Asuriya	Katabandh	18 (45.00%)	22 (55.00%)	18 (81.82%)	N.A.	4 (18.18%)
		Kotalpukur	17 (42.50%)	23 (57.50%)	19 (82.61%)	1 (4.35%)	3 (13.05%)
Cell Phone	**Barjora**	Paharpur	14 (35.00%)	26 (65.00%)	24 (92.31%)	2 (7.69%)	N.A.
		Mana	13 (32.50%)	27 (67.50%)	25 (92.59%)	1 (3.70%)	1 (3.70%)
	Hat-Asuriya	Katabandh	14 (35.00%)	26 (65.00%)	22 (84.62%)	N.A.	4 (15.39%)
		Kotalpukur	9 (22.50%)	31 (77.50%)	26 (83.87%)	N.A.	5 (15.13%)

Source: Field Survey 2012–13

potential of individual members, female participation in the diversified work/movement outside the home is also influenced by their socio-economic mobility/family sanction.

It is often observed that the hierarchical superiority of male-headed households over their womenfolk has a relatively constraining impact on the socio-economic mobility enjoyed by women, compared to those in female-headed households. In this context, the fact that mobility is a multidimensional issue should be borne in mind. It is shaped by gendered access to information, rights to land and other tangible assets, like money, education, skills, health status and patriarchal control. The daily mobility of women in developing countries is guided by a set of complex factors like prevalent social/cultural norms, transport infrastructure, geographical location, participation in various income-earning activities and SHGs, and access to information and communication technologies. However, like many other studies, we focus on the travel behaviour of women in different conditions of family guardianship.

Rural travel patterns in developing countries can be categorized into three broad types: (i) domestic travel, including water and firewood collection as well as food processing trips to grinding mills; (ii) agricultural travel, including trips to and from the fields; and (iii) travel for access to services and social purposes, particularly health facilities, shops, markets, temples/mosques etc. While travel burdens are often shared between men and women for agricultural purposes, women are almost entirely responsible for all domestic travel, which is by far the most energy- and time-consuming category in rural areas, accounting for one third to over two thirds of all travel (Peters 2001).

Table 6.6 below shows that both within and outside the village level, the perceived female mobility is higher in the case of female-headed households compared to male-headed households. This is the case in both the gram panchayats. Apart from being subdued by social taboos, females in male-headed households are less likely to own a vehicle or have a license to drive, are compelled to pursue a guarded movement with male possessiveness, and are less likely to have access to independent spending in the market. On these counts in female-headed households, circumstances drive women members to enjoy better freedom of movement and pursue independent choices in the market.

Table 6.6. Female mobility status on the basis of gendered head of the household

Gram Panchayat	Gender of Head of the Households	Region of Mobility	Mobility Status	No	Yes
Barjora	Female	Within village	Ability to move alone	22.22	77.78
			Family allowance to go out to market	22.22	77.78
			Family allowance to go to health facility	22.22	77.78
		Outside village	Family allowance to go to place outside this village/community	44.44	55.56
Barjora	Male	Within village	Ability to move alone	43.66	56.34
			Family allowance to go out to market	45.07	54.93
			Family allowance to go to health facility	39.44	60.56
		Outside village	Family allowance to go to place outside this village/community	59.15	40.85
Hat-Asuriya	Female	Within village	Ability to move alone	25.00	75.00
			Family allowance to go out to market	25.00	75.00
			Family allowance to go to health facility	0.00	100.00
		Outside village	Family allowance to go to place outside this village/community	25.00	75.00
Hat-Asuriya	Male	Within village	Ability to move alone	44.74	55.26
			Family allowance to go out to market	39.47	60.53
			Family allowance to go to health facility	31.58	68.42
		Outside village	Family allowance to go to place outside this village/community	60.53	39.47

Source: Field Survey 2012–13

8. Conclusion

The issue of diverse forms of unpaid female labour within the precincts of the familial atmosphere and the need for enhancing the scope and provision for extending their areas of participation in paid outdoor activities has long remained unrecognized and has been undervalued because of a lackadaisical attitude on the part of the relevant stakeholders. While many of them are overworked, both in housework and care for children and adults, and other work like tending to livestock and collection of common resources from the neighbourhood which are in general unpaid, their access to diverse types of income-generating activities remains constrained because of social taboos and lack of time, and no access to various forms of income yielding asset bases and labour market segmentation that tend to confine females to low returns and less productive job opportunities. The relatively poor literacy levels and fragile health tend to keep them unaware of their entitlements in the provision of various government policies, deter their interaction with bank officials, and render them unable to take part in better paid activities requiring physical strength. Again, the poor development of SHGs has stood in the way of micro credit-based earning opportunities of females. Inadequate opportunities to acquire skills and capital hinder their diversification prospects. Frequent NGO counselling, extension services, anganwadi visits and interaction with women would enhance their social interaction capabilities, raise awareness about the external world, and erase the taboos that they themselves nurture. Apart from this, the better implementation of midday meal schemes and provision for its monitoring with sudden checks/visits, a ban on child labour, improvement in the condition of primary health centres with specific provisions for women and the spread of health insurance facilities might help to generate the capabilities of women on the same level as their male counterparts.

References

Abdulai, A. & A. Crole Rees. "Determinants of Income Diversification Amongst Rural Households in Southern Mali." *Food Policy* 26 (2001): 437–452.

Bebbington, A. "Capitals and Capabilities: A Framework For Analyzing Peasant Viability, Rural Livelihoods and Poverty." *World Development* 27 (12) (1999): 2021–2044.

Dercon, S. & P. Krishnan. "Income Portfolios in Rural Ethiopia and Tanzania: Choices and Constraints." *Journal of Development Studies* 32 (6) (1996): 850–875.

Ellis, F. *Rural Livelihoods and Diversity in Developing Countries.*New York: Oxford University Press Inc., 2000.

Peters, D. "Gender and Transport in Less Developed Countries." *Gender Perspectives for Earth Summit 2002: Energy, Transport, Information for Decision-Making.* Berlin: Germany, 2000.

Quisumbing, Agnes R. "Household Decisions, Gender, and Development-A Synthesis of Recent Research." *International Food Policy Research Institute Washington, D.C.*, 1996.

Roemer, John E. "Economic Development as Opportunity Equalization." *Cowles Foundation for Research in Economics, Yale University*, 2006.

Smith, D. R. "Gender and the Rural Non-Farm Economy in Uganda." *Natural Resource Institute, Kent*, 2001.

World Bank. "Equity and Development." 2006.

CHAPTER SEVEN

FEMALE EMPOWERMENT THROUGH MICROFINANCE IN CRISIS: THE PHENOMENON OF MULTIPLE LENDING

SOUMITRA SARKAR

1. Introduction

Microfinance has been considered a double edged weapon in the fight against poverty. This movement started in India along with other countries of the world to achieve the Millennium Development Goal (MDG) and to give the impetus to the all-round development of the country. Microfinance includes micro-savings, microcredit, micro insurance, micro-pensions and micro-remittance. It provides low-cost finance and non-financial services to the vulnerable rural poor of the country to allow them to generate income through their microenterprises, which helps them improve their financial strength. The new paradigm of microfinance emphasises the economic and social empowerment of women through Self Help Groups (SHGs). It aims to improve the living standards of vulnerable rural women by empowering them in all respects and operates on the principle of "the borrower knows best." With this aim, the new microfinance program changes its face from an individual to group approach in terms of Self Help Group (SHG) in an effective manner. Harper (1998) stated that the group system of micro-financial services evolved in particular in Bangladesh (the Grameen Bank model) and in Latin America (solidarity groups and village banking), as well as in India (self-help groups). Sen (2003) identified that "the first and foremost element is self-help with the involvement of the poor in their own development preferably as a group. The collective coming together of individuals as groups commonly called self-help group sis the building block of microfinance in India."

The Self Help Groups provide the benefits of economies of scale, a cost effective alternative for different financial services, collective learning, a democratic and participatory culture, and a firm base and platform for dialogue and cooperation. Self-help implies a step from the stage of passivity to activity, and tomaking a creative contribution.

The National Bank for Agricultural and Rural Development (NABARD), in a pioneering effort, took the initiative in 1992 and issued operational guidelines to the commercial banks for a pilot project linking 255 Self Help Groups to the banks. It was an attempt to link the formal banking system to the informal groups in a cost-effective manner and to develop a transparent supplementary credit system for reaching the rural poor, with advantages to both them and the banks. Following the same mechanism, the Government of India formulated an umbrella scheme named Swarnajoyanti Gram Swarojgar Yojana (SGSY).

The microfinance movement, in the form of SGSY, was initiated in India on April 1, 1999. As an initial breakthrough, a total of 292,426 groups were formed under SGSY in 1999–2000. After 2011, more than ten million SHGs were formed under SGSY in India. The microfinance movement in all districts of West Bengal started after 1999 with the new mechanism, subsequent to the institution of the SGSY. A total of 337,499 SHGs were formed in West Bengal, and 22,742 Groups were formed in Jalpaiguri district under SGSY by the end of March 31, 2012.

Apart from government and development bank programmes, a large number of private microfinance institutions operates in India with different scales and motives. PMFIs, with their doorstep banking system, provide group-based loans to the vulnerable sections of society for different sophisticated terms and conditions which induce poor women to enter into different MFIs simultaneously. This phenomenon often provokes the beneficiaries to borrow from multiple institutions and enter into debt traps. When microfinance beneficiaries can access multiple MFIs, the tendency to borrow from all of them gradually increases, especially when the poor women find that the access to government programmes through banks is easily available, together with doorstep banking made available by private microfinance institutions. In these situation concerns are expressed about poor individuals' abilities to make wise financial decisions. Some research works are being taken up in this area, underscored by a belief that poor financial skills lead to poor decisions which ultimately lead to over indebtedness and debt traps. Aiyar (2009)noted that some women have "borrowed from four or more different MFIs, and in the process stepped into debt traps, which also hit MFIs through higher defaults." Recent concerns are therefore focused on the acute deficiency in making rational decisions relating to credit and debt alternatives. Bayang (2009) observed that: "worse incidents happen where micro finance beneficiaries borrow from one lender to pay other lenders. In the books of development agencies, it is still called repayment. When it comes to reporting during project evaluation stages, figures show that repayment is meeting recovery targets. What microfinance and development agencies fail to realise is that repayment of credit is not an indicator for the total success of the scheme. This study deals with the impact of microfinance on female empowerment in the phenomenon of multiple lending in the Rajganj block of the Jalpaiguri district.

2. Review of Literature

A significant number of studies have been carried out to assess the impact of microfinance on female empowerment. At first it will be prudent to throw some light on the concept and measurement of female empowerment. Kabeer (2001) stated that "empowerment" implies increased participation in decision-making, and it is this process through which people feel capable of making decisions and realize their right to do so. The World Bank Report (2001) identified four key elements of empowerment to draft institutional reforms: access to information; inclusion and participation; accountability; and local organisational capacity. Hashemi, Schuler & Riley (1996) investigated the change in female empowerment with the help of an ethnographic study and quantitative survey. They create an empowerment indicator built on eight criteria: mobility, economic security, ability to make small purchases, large purchases, involvement in major household decisions, and relative freedom from domination by the family, political and legal awareness, participation in public protests and political campaigns. Previously, the repayment rate was used as an indicator of empowerment.

Nagayya (2000) maintains that an informal arrangement for credit supply to the poor through SHGs is fast emerging as a promising tool for promoting income-generating enterprises. Ahmad (1999), through a case study on thrift groups in Assam, highlighted that women are coming to the administration directly for their just rights and to address their grievances boldly. This proved that self-help groups are successful in North East India, even in the midst of insurgency. Similarly, Gurumoorthy (2000) mentioned that the female led SHGs have successfully demonstrated how to mobilize and manage thrift, appraise credit needs, maintain linkages with the banks and enforce financial self-discipline. SHGs enhance the equality of status of women as participants, decision-makers and beneficiaries in the democratic, economic and social and cultural spheres of life, and encourage women to take active part in the socio-economic progress of the society. Bhatia & Bhatia (2000), through some case studies, highlighted that recovery of SHGs is higher than other credit extended to borrowers, and they observed that there have been perceptible changes in the living standards of the SHG members in terms of ownership of assets, increase in saving and borrowing capacity, income generating activities and income levels as well. Rao (2002) identified that a review of the genesis and development of SHGs in India reveal that the existing formal financial institutions have failed to provide finance to landless, marginalized and disadvantaged groups. Puhazhendhi (1999) observed that SHGs in Tamil Nadu had been performing well regarding social change and transformation. The emerging trends are leading to the positive direction of empowerment of members and promotion of microfinance. Dasgupta (2000) viewed microfinancing through the informal group approach, giving rise to several benefits: savings mobilized by the poor; access to the required amount of appropriate credit by the poor; matching the demand and supply of credit structure and opening new market for financial institutions (FIs); reduction in transaction costs for both lenders and borrowers; tremendous improvement in recovery; heralding a new realization of subsidy-less and corruption-less credit, and the remarkable empowerment of poor women. Barbara & Mahanta (2001) maintained that the SHGs have helped to set up a number of micro-enterprises for income generation. Puhazhendhi & Satyasai (2001) viewed that the SHGs as an institutional arrangement could positively contribute to the economic and social empowerment of the rural poor. Manimekalai & Rajeshwari (2001) highlighted that the provision of microfinance by the NGOs to female SHGs has helped the groups achieve a measure of economic and social empowerment. It has developed a sense of leadership, organizational skill, management of various activities of business, the right to acquire finance, identification of raw material, marketing and choice of suitable diversification, and modernization strategies. Similarly, Sharma (2001) maintained that through SHGs female empowerment had been taking place. Their participation in the economic activities and decision-making at the household and society level is increasing and making the process of rural development participatory, democratic, sustainable and independent of subsidy; thus, micro-financing through SHGs is contributing to the development of rural people in a meaningful manner. Interestingly, Singh (2001), in his study in Uttar Pradesh, highlighted that the SHGs were functioning in place of moneylenders because loans could be taken at any time as and when needed for any purpose. There are no formalities involved and the transaction cost is low. Sarkar (2006) found that female micro entrepreneurs in North Bengal in West Bengal had profitably taken up non-traditional economic activities, which developed positive thoughts in the minds of the women and increased their economic benefits. Sarkar & Dhar (2009) argued that the economic and social status of the female beneficiaries had increased to a significant extent after joining an SHG.

Zaman (2001) argued that greater access to resources via micro-credit enhanced female control over her assets. The control which a woman has over her assets is measured by, amongst other things, her ability and right to sell assets on the basis of her own personal choice. Carloni (1987) found that credit programmes were more successful than income-generating projects in having a genuine impact on women's economic statuses.

Amin & Pebley (1994) suggested that BRAC's (Bangladesh Rural Advancement Committee) loans contributed to an increase in a woman's household decision-making power, her control over household resources and physical mobility outside the home. It is interesting to state that women believed that receiving microcredit loans reduced their chances of abandonment. Their husbands regarded these women as a valuable resource. Banu et al. (2001) found that BRAC had been able to bring about considerable changes in the lives of its female beneficiaries by facilitating their material, perceptual and relational pathways to empowerment, both at the individual and households levels. Banu et al. (2001) also concluded that husbands learned to value women more because they were now perceived to be an avenue for loans for capital investments by the household. Carr, Chen & Jhabvala (1996) argued that the lives of women changed partly because microcredit loans took them outside the household and required their participation in BRAC village organizations. Thus, women developed links to the wider community group. In another study of impact assessment, Hashemi, Schuler & Riley (1996) claimed that involvement in the Grameen Bank's or BRAC's credit programmes certainly empowered women by increasing their physical mobility, their ability to make purchases and major household decisions, their ownership of productive assets, their legal and political awareness, and participation in public campaigns and protest. This study also pointed out that the incidence of violence against women was reduced as aresult of female involvement in microcredit programmes.

It is evident from the literature that a paradigm shift in the microfinance programme has a great significance for the development of the SHG beneficiaries. Interestingly, several studies have been conducted by social scientists, financial institutions and agencies which highlight the positive trends and impacts of self-help groups on empowerment, credit accessibility and social change. It is very difficult to review all the relevant studies since the proper documentation of such studies has not yet been ensured. Therefore, the available relevant studies, particularly case studies, workshops, seminars and symposia, have been critically reviewed.

The above studies simply demonstrate that SHGs are playing a vital role in extending micro-finance to the rural poor. The functioning of SHGs has been based on participatory mechanism and therefore the impacts of SHGs on its members in terms of empowerment, accessibility to credit, socio-economic change etc. has been positive. Though there are a number of studies related to the functioning of micro-finance, only a few studies assess the impact of female self-help groups on socio-economic empowerment. However, many researchers have expressed concern about this, pointing out that women may repay loans through taking them elsewhere, getting into serious debt. In some cases men may control loans, whereas women might be mediating between male family members and MFIs. On measuring empowerment, Kabeer (1999) highlighted various methodological points about some of the key elements like resources, agency and achievement. Many analysts pointed out the need to go beyond "access" indicators to grasp how "resources" translate into the realization of choice, and that they have led to a variety of concepts seeking to bridge the gap between formal and effective entitlement to resources by introducing some aspect of agency into the measure. "Control" is one of the most commonly used ways to measure empowerment. For instance, Goetz & Sen Gupta (1996) argue that if control over loan-driven activity is in fact a critical "control" point in the process of access to loans, translating into a range of valued achievements, then "managerial control" can serve as an indicator of empowerment. Malhotra (2002) emphasised that even after identifying empowerment as a primary development goal, neither the World Bank nor any other major development agency developed a rigorous method for measuring and tracking changes in the levels of empowerment.

In this context, the present study is important for assessing the impact of female self help groups on the economic empowerment of the members in the Rajganj block to find out whether multiple lending has any untoward impact on empowerment of female beneficiaries. The study findings may be useful for policy implications and the smooth functioning of SHGs.

3. Objective of the Study

A large body of literature expresses that there are a sizeable number of research studies on the Microfinance Programmes and SHG movement in different countries. Many studies concentrate on diverse topics in the areas of problems related to the empowerment of women through microfinance schemes. However, except for one or two studies, no focus has been made particularly on the problems of empowerment of SHG women in West Bengal, ona holistic basis. Further, holistic studies on microfinance in North Bengal have not yet been carried. This study, focussing on one district in North Bengal, can be seen as a filler in this respect.

The main objective of the study is to make a comparative impact assessment of SHGs involved in multiple borrowing and those not involved in multiple borrowing to find out the impact of microfinance interventions on the economic empowerment of the female beneficiaries.

4. Methodology of the study

4.1 Sample

A survey was carried out with the SGSY beneficiaries as the target respondents to find the answer to these questions. Before any questionnaire was administered, detailed discussions were held with some of the SGSY beneficiaries living in the areas and a pilot survey was carried out. The final structured questionnaire was formulated on the basis of preliminary interviews and pilot surveys. The questionnaire was administered to a random sample of 476 SHGs in the Rajganj block under the Jalpaiguri district in North Bengal. This study divided the population into two different categories: the SHGs who were not involved in multiple borrowing, and the other category of SHGs who were involved in multiple borrowing. It implies that each sample group consists of 238 SHGs. The target respondents were the only SHGs under the SGSY who had passed grade one and credit linked with banks. The population of such groups in the area under study was more than 1,318, and the sample therefore represented more than 36.11% of the population. Through long discussion about the socio-economic problems of the beneficiaries, particularly those involved in multiple credits, different particular cases were identified. The findings of the study are presented in the next section.

4.2 Methodology of impact measurement

A large body of literature evidences that impact assessment studies have mostly used a questionnaire supplemented by case studies. While a structured questionnaire helps in the objective assessment of the percolation and perception of the benefits of microfinance on an aggregate basis, case studies focus on the individual/group level impact on socio-economic uplift. The findings from the amalgam of these two techniques provide an understanding of the impacts at the grass-root level and help in restructuring the microfinance programmes.

5. Findings of the study

A group-based modern microfinance programme was implemented in the catchment area of the study in 1999, and became fully fledged during 2001–02 in most of the GPs of this block. The level of changes made by group-based microfinance during its decade-long journey in this district will be assessed in the following section. Impact assessment will be described with the help of an index of benefits that a beneficiary can enjoy from the modern microfinance provision.

The most obvious benefit would be the cash income accrued, but it may also be necessary to consider other economic benefits such as access to loans, regularity of work received and change in asset possession. The perception of such economic benefits, along with the tangible benefits, may differ amongst members depending on the socio-economic status of this block. Assessment of all the above factors would provide a more clear picture about the economic benefits accrued from the income-generating activities rather than considering only the income earned. An index of economic benefits has been created with the help of the aforesaid six variables. The actual and perceived economic benefits are accrued to the bonafide members of this block. A higher level of economic benefits enjoyed by the members is identified by a higher value of index, while a lower value indicates the lesser economic benefits received.

The findings of the study are reported in the following sequence. First, a detailed distribution of each of the factors considered for the formulation of the indices has been presented for each of the blocks. The factors have then been taken to formulate and report on the economic and social indices for each block and for the district as a whole.

5.1 Regularity of work

Economic benefits accrued or earned by SHG members from the micro-enterprises run by them depend to a large extent on the frequency of work done in these enterprises. Some of the micro-enterprises are occasional in nature, in which the members are engaged in earning seasonal income. Many other micro-enterprises operate throughout the year in which SHG members are engaged over the entire year on a regular basis. Regularity of work depends on the nature of micro-enterprises run by the members.

Table 7.1. Distribution of SHGs on the Basis of Regularity of Work

Regularity of Work	SHGs who are not involved in borrowing from other institutions	SHGs involved in multiple borrowings
	%of total	% of total
Every day	45.00	35.00
5 day in a week	50.00	0.00
3–4 days in a week	5.00	16.25
1–2 days in a week	0.00	28.75
No Work	0.00	20.00

Source: Field Survey

Table 7.1 shows that the highest percentage (45%) of respondent SHGs, who are not involved in borrowing from other institutions, received work from their micro-enterprises every day, whereas 35% are involved in multiple borrowing received every day. It is important to note that, out of the SHG members involved in multiple borrowing, 20% did not start any micro-enterprises which give regular work to the beneficiaries. One of the main reasons behind this is that most of the members are involved in debt swapping, so they could hardly accumulate the necessary funds to start the micro-enterprises.

5.2 Average annual income of the groups

The cash income of a group is generated from different sources, and a good source of regular income is interest on group savings. Another component of group income is interest on inter-loans among the members within the group. Income from group economic activities forms the main component of group income. In cases where the SHG members individually operate separate micro-enterprises with the financial assistance from the group, the income of such micro-enterprises has also been considered for the calculation of the average annual income of the group. The financial management of the group corpus has a significant impact on the income generation of the group. The average annual income of the group differs on the basis of the nature of economic activities taken up. As stated earlier, the nature of economic activities depends on various crucial factors. Basically, the temporary nature of micro-enterprises, like the seasonal business of agricultural products, the development and maintenance of picnic spots, and mushroom cultivation, is unable to provide a steady flow of income for the whole year. As a result, the average annual income of most of the groups is not substantial. The average annual income level of the respondent groups in the catchment area of the study is given in the following table.

Table 7.2. Distribution of SHGs on the basis of average annual income of the group

Average Annual income of the SHGs	SHGs who are not involved in borrowing from other institutions	SHGs involved in multiple borrowings
	%of total	% of total
More than Rs. 20,000	20.00	15.00
Rs. 10,001–Rs. 2,000	10.00	10.00
Rs. 5,001–Rs. 10,000	60.00	40.00
Less than Rs. 5,000	10.00	35.00

Source: Field Survey

Table 7.2 shows that the average annual income of the beneficiaries in most cases belongs to the range below Rs. 10,000, though the proportion is higher in the case of the SHGs under SGSY who are not involved in multiple borrowing. In comparison, the percentage of the SHGs involved in multiple lending and unable to generate an average annual income more than Rs. 5,000 happened to be the highest. A significant reason behind it is that the lion's share of their generated funds is being utilised to repay the loan of the PMFI.

5.3 Actual monthly income of the members

A fully-fledged SHG based microfinance movement started in the Rajganj block during 2001–02. A substantial portion of the total number of SHGs in this block have passed the 1st grading and are engaged in traditional micro-enterprises like tailoring, chira muri (thrashed rice)making, wool knitting, goatery, piggery etc. Even in most of the cases, members have started these types of micro-enterprises as sole proprietorship businesses after taking financial assistance from the group. So naturally, the average monthly income of the members is not substantial, and differs on the basis of types of micro-enterprises undertaken and the other economic opportunities available. The category wise distribution of the SHGs on the basis of actual income earned by the members is given in the following table.

Table 7.3. Distribution of SHGs on the basis of actual income earned by the members

Average monthly income of the members	SHGs who are not involved in borrowing from other Institutions	SHGs involved in multiple borrowings
	%of total	% of total
More than Rs. 2,000	0.00	0.00
Rs. 1,001–Rs. 2,000	10.00	0.00
Rs. 500–Rs. 1,000	25.00	12.50
Less than Rs. 500	65.00	87.50

Source: Field Survey

Table 7.3 shows that the average annual income of 65% of respondents of SGSY only in this block is less than Rs. 500. On the other hand, 87.5% of the respondents involved in multiple borrowing were able to earn less than Rs. 500 per month. It is interesting to note that 10% of the respondent SGSY beneficiaries not involved in multiple borrowing are able to generate income in the range of Rs. 1,000 to Rs. 2,000, whereas no SHGs involved in multiple borrowing were able to generate income in the aforesaid range. The main reason behind this is that in most of the cases sample SHGs have taken loans from different PMFIs at the same time, and cannot manage it.

5.4 Access to loans

Loans are an essential component of any economic system which motivates people to start enterprises, a vital source of finance though which an entrepreneur can raise the funds to substitute the capital requirement of the enterprise. The new paradigm of microfinance provides multilevel loan facilities to the beneficiaries in order to start and operate vibrant micro-enterprises for them. The degree to which SHGs help in enhancing the members' access to loans is symptomatic of the degree of economic benefits provided by that group to its members. Basically, rural women are unaware of the formal banking system. They have been treated as a bank-illiterate section of the society and also find themselves increasingly marginalised from access to formal credit on account of a variety of socio-cultural biases and bureaucratic processes. The self-help group provides inter-loaning facilities to its members to meet their emergent credit needs. Banks also provide collateral free loans at low costs to the beneficiaries to start and operate their micro-enterprises. It is interesting to note that most of the members have taken loans for the purpose of day-to-day survival, initially from local moneylenders at exorbitant interest rates, and this has further pushed the poor into a vicious cycle of poverty and debt. This trend has improved gradually with the passing of time. Modern microfinance provides multilevel financing to inject the earning potentiality of the beneficiaries. Table 7.4 below shows the amount of loans taken by the beneficiaries of this block in Jalpaiguri district.

Table 7.4. Percentage distribution of members by access to loans

Access to loans through group	SHGs who are not involved in borrowing from other Institutions	SHGs involved in multiple borrowings
	%of total	% of total
More than Rs. 20,000	0.00	0.00
Rs. 10,001–Rs. 20,000	0.00	18.75
Rs. 10,000–Rs. 5,001	25.00	25.00
Less than Rs. 5,000	75.00	56.25
Zero Amount	0.00	0.00

Source: Field Survey

The study shows that most members have taken loans on an average less than Rs. 5,000. Promptness of loan disbursement of PMFIs is the centre of attraction for the poor women as the beneficiaries often feel the need for funds. The PMFIs have been observed to issue loans of smaller amounts as much as possible.

5.5 Perception of economic benefits

Modern microfinance provides financial and non-financial services to support income generation activities with the goal of enhancing the economic status of the beneficiaries. The success of such an endeavour is assessed on the basis of number of beneficiaries involved and income earned by them. The main aspect of such evaluation consists in assessing how members perceive those economic benefits traditionally considered as quantifiable cash rewards. On the basis of individuality and their level of aspiration, it was realised that the same cash benefits could provide different perceptions of well-being. In order to bring this factor into impact assessment, individual realisations about the economic benefits accrued from the SHGs have been taken into consideration. Different levels of perception about the economic benefits for the well-being of the beneficiaries of the Rajganj block in Jalpaiguri district are detailed in Table 7.5 below.

Table 7.5. Percentage distribution of members on the basis of perception of improvement in economic status

Perception about improvement in economic status	SHGs who are not involved in borrowing from other institutions (% of total)	SHGs involved in multiple borrowings (%of total)
Much Better	30.00	10.50
Somewhat better	65.00	65.75
Remained the same	5.00	12.75
Decreased	0.00	11.00

Source: Field Survey

Table 7.5 shows that the beneficiaries of this block perceived their economic status to have increased after joining the SHG, and that self-help groups provide economic benefits to them. After the thirteenth year of SHG movement in this district, the economic status in this block increased to some extent. It is interesting to note that there are a few members who felt that the economic benefit had decreased due to multiple borrowing. Most of the beneficiaries in this block realized a moderate level of improvement in economic status.

5.6 Change in asset possession

It is a general phenomenon that the assets of rural families are usually owned in the name of the male members. In many cases, rural women purchase domestic animals like cows, goats and poultry to earn their livelihoods. Rural women usually use their earnings to supplement the consumption expenses of their families. In some cases, those female members purchased assets like jewellery, televisions and bicycles for personal purposes with their surplus funds after meeting the necessary consumption expenses.

The field survey of this study reveals that female members of some SHGs in this block were able to buy different assets in small amounts. Taposhi Roy, a member of the Tagar Self Help Group in this block, stated that after joining the SHG she was able to purchase one cow and one bicycle for her family. Asset purchasing power differs on the basis of the consumption needs of the families. The extent of this need also depends upon the earning capacity of the males. Where males earn a minimum amount as daily wages, females have to share the financial responsibility to run their families, otherwise she could spend her limited earnings purchasing the necessary assets for her family. The changes in asset position of members in this block after joining the SHGs are presented in Table 7.6 below.

Table 7.6. Change in value of assets in post SHG period

Change in assets value	SHGs who are not involved in borrowing from other Institutions	SHGs involved in multiple borrowings
	%of total	% of total
Increased	80.00	45.00
Remained same	20.00	40.00
Decreased	0.00	15.00

Source: Field Survey

Table 7.6 above shows that SHGs have a significant impact on the asset position of members' families. In a few cases, members were not able to bring any asset to their families after joining the SHGs. There are two possible reasons behind this situation: firstly, male members were unable to run their families due to low incomes and female members had to take up the responsibility. In these cases, they spent their limited earnings from group activities fulfilling the emergent consumption needs. Secondly, members were often not able to run their productive group activities efficiently, which in turn stood in the way of earning any substantial income for them. As a result, there was no question of their purchasing any assets. It is interesting to note that there are some members involved in multiple borrowing and whose asset position decreased after joining the SHG.

This study has found that the economic empowerment of women is somewhat affected if the beneficiaries' accessing loans from multiple sources.

This study also delves into multiple borrowing through the case vignettes formulated on the basis of personal interviews with the beneficiaries on several occasions. Several cases of heterogeneous natures were found in the course of this study. Two interesting cases of multiple borrowing are given in the following section.

Case-I:

Ma Durga SHG was formed in 2002 by the combined effort of block and DRDC officials involving thirteen homemakers of marginal farmers of the Panikouri Gram Panchayat under the Rajganj block in Jalpaiguri district. The pooled savings of this group were linked with the United Bank of India (UBI) Panikouri branch. After the successful completion of one year and two months, they passed their first grading test and received Rs. 15,000 as revolving funds from DRDC. The bank released Rs. 20,000 as an advance to the group. Suniti Sarkar, one of the needy members, applied for a loan from the group. The group members unanimously sanctioned the loan to the applicant to the tune of Rs 10,000. She gave this to her husband to open a small grocery shop at her small hut. Initially, the small but steady flow of profit from the business proved its financial viability and it became a worthwhile enterprise. Subsequently, she started to repay her group loan from this profit.

After repayment of 50% of the loan amount, her husband decided to plough the profit back in to enhance the volume of his business, while she felt that there were insufficient funds for such an improvement. She could not arrange the required funds from her internal source and was unable to run the business properly. Seeking help, she contacted a member of a group operated by a private microfinance institution in a nearby village, asking the field officer of that MFI to disburse a loan of Rs. 25,000 to her. As per the consent of all the members, who were present in their weekly group meeting, the field officer of PMFI collected the loan application form and gave permission to the group leader to register her name as a member of that group. She had to submit Rs. 250 as a loan processing charge and Rs. 100 as an insurance premium as preconditions for this new loan. Finally, Suniti Sarkar received Rs. 23,000 in cash and as per the business policy of this MFI the difference, Rs. 2,000, was kept by the officer as collateral. He promised that it would be adjusted in time for the final repayment of this loan.

Loan recovery started in accordance with the recovery schedule of this MFI, which was a weekly instalment system, immediately after the first week of loan disbursement. The beneficiary had to pay Rs. 575 per week, and it was supposed to be completed by 52 weeks. It was a great financial problem for her to arrange the requisite fund to repay two loans at a time. She had to convince other members of the SHG to allow her to postpone the loan repayment, and promised that she would later repay the loan with adequate interest. On the other hand, it was impossible for her to arrange the required funds for the weekly instalment of Rs. 575, and ultimately she could not pay her second and remaining instalment. Other members of the group had arranged the amount of the second and third instalments. It is the collection policy of the private MFI that the loan taker has to repay the loan on time by any means without failure. The field officer would not compromise with default, and could not leave the group meeting without the full recovery of the total weekly instalment amount. All group members exercised peer pressure to recover the remaining amount from Suniti Sarkar and threatened to dispose of her valuable property to recover the balance amount of the loan. To end this hardship, she had to gradually close down the grocery business and open a tiny betel leaf stall at a corner of her hut to sell a part the grocery stock. Members of both the groups had pressurized her to repay the loan with adequate interest, and as a result she was not able to run the betel leaf stall for long.

She suffered from a lack of adequate money and searched for another loan to repay the previous loan from PMFI. At last she was able to convince the field officer of another private MFI of the urgency of the loan by suppressing the original facts, and he gave her permission to register herself as a member of an existing group. The field officer sanctioned Rs. 15,000 as a loan from this MFI after the payment of Rs. 150 as a loan processing charge and Rs. 100 as an insurance premium of this loan. She received Rs. 13,500 in cash from this institution, though the sanctioned amount was Rs. 15,000. The balance of Rs. 150 of the third loan would be adjusted with the last few instalments. She had used this third loan amount for part payment of the previous loan from the private MFIs. Here, she had to pay the weekly instalment of Rs. 340 for fifty weeks commencing immediately after the first week of loan disbursement. After the third loan, the total outstanding loan amounted to Rs. 38,475, for which she had to pay Rs. 915 as a weekly instalment to the MFIs.

Gradually, she was unable to repay any instalment at all and had to arrange for the required funds from her nearest relatives for the first two instalments of the second MFI. She had to sel lher cow for Rs. 5,000 and gave it to the first MFI towards repayment of the second loan. Afterwards, she also had to close her betel leaf stall and took another mortgage from a local moneylender, keeping her small hut as collateral to repay the total outstanding loan amount. Over time, she became overwhelmed by a debt trap.

Case-II:

Putul Roy is a homemaker of a poor family. Her husband is a daily labourer, working on a daily wage basis under a building contractor. She has one son and two daughters. She lives in great privation with very little resources for her daily existence. To come out from this condition and to help her family, she was inspired by the local panchayat and motivated by the neighbours to form a self-help group (SHG). Finally, she managed to form an SHG with twelve other women in the locality. Their savings were linked with UBKGB, Fulbari. She was facing the financial impossibility of continuing the education of her three children. In order to meet this financial need, she took a loan of Rs. 4,500 from the group.

After their group passed the first grading, the DRDC gave them Rs. 10 000 as a revolving fund and the bank sanctioned Rs. 15,000 as a loan to start a new joint venture for their group. Initially, they were able to start a group-based joint venture and distributed the funds among the members as per their needs. Naturally, she got a loan from the group to the tune of Rs. 7,000 and started a tailoring shop in the locality. She purchased a large volume of cloth from local market on a cash basis and began to supply the garments in the local market on credit. It took a long time to realise the fund from the purchaser, and there was a delay in getting the funds from her customers. As a result, there was a need of more funds to manage the enterprise efficiently. In addition, she was searching for the same from some outside sources since it was impossible for the group to supply the fund due to non-repayment of the previous loan, which is one of the main conditions of the SGSY. At this juncture she heard news from the nearby villagers that a private microfinance institution was giving loans on easy terms and in the shortest possible time. She went to the nearby village to meet the supervisor of a private microfinance institution to discuss the opening of a new group. Finally, she was able to form a group with twenty-four other women in the neighbourhood. Thereafter, she gave the application to the supervisor through the group leader. After one week of submission of loan application, the supervisor disbursed the loan at the doorstep of the group. She succeeded in getting a loan of Rs. 20,000 from the PMFI to manage her enterprise efficiently and to earn an adequate profit.

It took a long time to run the enterprise in a better way with adequate funds. During that period the amount of the first loan, the principal and interest altogether, mounted to a large volume. The peer group had put pressure on her to repay the loan somehow. Naturally, a large part of the last loan amount had been used to repay the first two loans of SHG under SGSY. Again, she fell into a financial crunch in meeting the need corresponding to the second loan taken from the PMFI. On the other hand, as per the rules of the PMFI, the borrower must start their first instalment of repayment immediately after a week of the loan disbursement. The collection machinery of the PMFI is so effective that there is no scope of avoidance of any instalment. As a result, she had to repay two instalments at a time, and it was very difficult for her to arrange the funds simultaneously and she had to sell her small house and a plot of acquired land to do so. Sometime afterwards they became homeless and had to stay at their relative's house on a temporary basis. They could hardly find way to merely survive with dignity.

The facts revealed by the beneficiaries demonstrate that debt swapping is indeed a serious phenomenon which has proliferated subsequent to the emergence of microfinance systems in the area under the study. Excess debt intake often results from the absence of adequate knowledge of the debt schemes, i.e. the term of debt, conditions of debt, rates of interest etc. Here lies the importance of the phenomenon of financial literacy, which is a pre-requisite for handling of funds from multiple sources, and "debt literacy" is an important component of overall financial literacy. Lusardi & Tufano (2009) defined debt literacy as: "the ability to make simple decisions regarding debt contracts, in particular basic knowledge about interest and compounding, measured in the context of everyday financial choices."

6. Conclusion

There is no doubt that microfinance is applauded and the beneficiaries can reap its benefit if, and only if, they can operate the income generating micro-enterprises in a sustainable manner for a prolonged period. The status of economic attributes reflects the level of economic empowerment. The study has significant findings that some women beneficiaries are involved in multiple borrowing, meaning that microfinance beneficiaries under the SGSY scheme have also been accessing loans from PMFIs. The beneficiaries considered that it is relatively easier to access loans from

PMFIs than government sponsored microfinance schemes. The access to simultaneous loans creates serious problems for the vulnerable and financially illiterate rural women. In most cases, they are unable to meet the debt obligation as per the schedule which leads to losing their valuable assets, even their houses. This common phenomenon of over indebtedness and debt swapping of the female beneficiaries compels them into debt traps. Therefore, it can hardly be said that microfinance has a necessarily significant impact on the effective and real empowerment of female beneficiaries, and in many cases the important factors of female empowerment have been moving in the wrong direction.

To control the situation, as a remedial measure the Government of India (GOI) passed the Microfinance Regulation Bill in 2012. The provisions of the said act should be strictly implemented at the grass roots level so that multiple lending can be put to an end on a permanent basis. As an advance step, GOI has reformulated the SGSY scheme, which is termed the NRLM (National Rural Livelihood Mission). All the lacunae of the previous programme have been taken into consideration in the restructuring of the new microfinance programme at the government level. Proper implementation of the NRLM for training, development and identification of productive activities can enhance the scope of the establishment of effective micro-enterprises. If this phenomenon of multiple lending cannot be stopped, the beneficiaries would not be able to reap the benefits of the microfinance programme and the crisis factors of female empowerment will continue and even lead to the disempowerment of some sections of womenfolk.

References

Ahmad, M. A. "Women Empowerment: Self Help Groups." *Kurukshetra*, April, 1999.

Amin, S. & A. R. Pebley. "Gender Inequality within Households: The Impact of a Women's Development Program in 36 Bangladeshi Villages." *Bangladesh Development Studies* 22 (2–3) (1994): 121–154.

Aiyar, S "How Micro-Finance Institutions Beat Nationalized banks." *The Times of India 26 July 2009*

Barbara, S. & Mahanta, R. "Micro Finance Through Self Help Groups And Its impact: A Case Of Rashtriya Gramina Vikas Nidhi—Credit And Saving Programme In Assam." *Indian Journal Of Agricultural Economics* 56 (3) (2001): July–Sept.

Bayang, S. "Where Micro Finance Fails to Meet Enterprise Goals." *Allgambian.net* Feb 1,2009

Bhatia, N. & Bhatia, A "Lending To Groups." Yojana, Feb., 2002.

Carr, M. M. Chen & R. Jhabvala (eds.). "Speaking Out: Women's Economic Empowerment in South Asia." *New Delhi: Vistaar Publications.* 1997

Dasgupta, R "An Informal Journey through SHG's." *Indian Journal & Agricultural Economics* 56 (3) (2001): July–Sept.

Dhar, S. N. & S. Sarkar. "Financial Sustainability of Microfinance Units Emperical Evidences From SHGs In India." *Indian Accounting Review* 13 (1) (2009).

Goetz, Anne Marie & Rina Sen Gupta. "Who Takes the Credit? Gender, Power, and Control over Loan Use in Rural Credit Programs in Bangladesh." *World Development* 24 (1) (1996): 45–63.

Gurumoorthy, T. R. "Self Help Groups Empower Rural Women." *Kurukshetra,* Feb., 2000.

Hashemi, S. M., Schuler, S. R. & Riley, A. P. "Rural Credit Programmes and Women's Empowerment in Bangladesh." *World Development* 24 (4) (1996): 635–653.

Kabeer, N. "Conflicts Over Credit: Re-Evaluating the Empowerment Potential of Loans to Women in Rural Bangladesh." *World Development* 29 (1999).

—. "Reflections on the Measurement of Women's Empowerment: In Discussing Women's Empowerment-Theory and Practice." *Sida Studies* 3. Novum Grafiska AB: Stockholm. 2001

Lusardi, Annamaria & Peter Tufano. "Debt Literacy, Financial Experiences, And Over Indebtedness." *Working Paper 14808, National Bureau of Economic Research, Cambridge, MA 02138* March, 2009.

Malhotra, R. "Access To Rural Women To Institutional Credit: Issues And Alternatives." *BIRD, Lucknow.* 2002.

Manimekalai, M. & Rajeshwari, G. "Nature and Performance Of Informal Self Help Groups—A Case From Tamil Nadu." *Indian Journal Of Agricultural Economics* 56 (3) (2001).

Nagayya, D. "Micro-Finance For Self Help Groups."*Kurukshetra*, August, 2000.

Puhazhehdhi, V. "Evaluation Study Of SHG's: Important Findings of Evaluation Study In Tamil Nadu." *Paper Presented In A Workshop, Dated* August 26–27, 1999, BIRD, Lucknow.

Puhazhendi, V. & Satyasai, K. J. S. "Economic and Social Empowerment of Rural Poor Through SHGs." *Indian Journal of Agricultural Economics* 56 (3) (2001): July–Sept.

Rao, V. M . "Women Self Help Groups: Profiles From Andhra Pradesh And Karnataka." *Kurukshetra,* April, 2002.

Sarkar, S. "Self Help Groups and Non Traditional Products: A Study of Micro Enterprises in North Bengal." *Indian Journal of Millennium Developments studies: An International Journal* 1 (1) (2006).

Schuler, S. R., S. M. Hashemi, A. P. Riley & A. Akhter. "Credit Programs, Patriarchy and Men's Violence against Women in Rural Bangladesh." *Social Science and Medicine* 43 (12) (1996): 1729–42.

Sen, A. "Gender and Cooperative Conflicts." *In Persistent Inequalities: Women and World Development,* ed. I. Tinker. New York: Oxford University Press, 2003.

Sharma: K. C. "Micro Financing Through SHG's." *Indian Journal of Agricultural Economics* 56 (3) (2001): July–Sept.

World Bank. "Engendering Development: Through Gender Equality in Rights, Resources, and Voice—Summary." 2001, www.worldbank.org/gender/prr/.

Zaman, H. "Assessing the Poverty and Vulnerability Impact of Micro-Credit in Bangladesh: a Case Study of BRAC." Unpublished background paper for World Bank, World Development Report 2000/2001 (Washington, World Bank).

SECTION C:

SOCIO-DEVELOPMENTAL ISSUES

CHAPTER EIGHT

HOW SUCCESSFUL IS INDIA'S LOOK EAST POLICY UNDER GLOBALISATION?

UTPAL KUMAR DE

1. Background

The integration of economies encouraged by the relaxation of trade and other barriers is supposed to create an atmosphere that would promote trade, investment opportunities for the companies beyond national boundaries, and accelerate capital flow in all forms including human, natural, manmade and other productive forms including technology, and thus lead to an accelerated economic growth of the countries that come closer in all such terms. The opening of countries and integration with the global economies are thus conceived to be the solution to several social and economic problems (World Bank & IMF 2007; Dreher 2006; Kulkarni 2005; Amavilah 2009a). The human development index is found to depend on some conventional factors and forces, national symbols and also significantly on global connections and interactions (Amavilah 2009b; De & Pal 2011). In the present global set up, free trade not only increases efficiency but also helps to reduce pollution emissions due to greater competitive pressure and greater access to greener production technologies and knowledge as well as international capital transactions (Cole 2004; Antiweiler et al. 2001). However, looking at the persistent unjust disparities between the developed and developing world, some doubts have been raised by Heintz (2006) and Cherni (2001) about the role of globalisation in meeting employment challenges and ensuring quality of working life, reduction in poverty and maintaining gender equity.

The globalisation process provides the scope to nations and their industrial and commercial establishments to participate in the international market efficiently and facilitate the flow of capital and other resources globally to accelerate development. In addition, the process brought a number of problems in the participating countries. The environmental problems that have occurred or worsened as a result of globalization in cities of the developing world can be added to a list of already existing and perhaps critical problems, such as poverty and lack of sanitation and running waste and accumulating urban waste (Cherni 2001). Thus, there are contradictory forecasts with regards to the impact of integration of economies on development prospects, as perceived by various researchers. Several limitations have been pointed out by a number of studies that warn against unrestricted liberalisation and opening of countries to the outside world without considering the competing ability of the domestic sectors (Beams 2000; Effland et al. 2006; Heintz 2006). Despite such caution, a large number of countries have followed this path blindly by bringing in foreign capital without considering its social, economic, demographic and environmental consequences (Tang 2008; Versi 2004; Ewege 2005). It is argued in line with trade theories that it is very difficult to progress beyond a certain point with a countries own efforts due to the lack of complete knowledge, technology and hence efficiency on many fronts, and thus interdependence and free trade have no other alternative and may lead to specialisation and accelerate trade that will mutually benefit the participating countries, leading to the faster growth of their economies. Of course, a few countries, despite being the signatory to many international treaties, follow the path of globalisation and open economy. Along with that, those countries apply some restrictions to safeguard the interests of various domestic industries, their employees and markets, and also to protect their socio-cultural values.

India's Look East Policy (LEP) is a policy towards integration with the eastern neighbours for the growth of trade and thereby accelerates the growth process. After becoming a fully-fledged member of ASEAN in 1995, in 2002 India entered into annual dialogue process for the promotion of trade, socio-cultural exchange, technological cooperation, resource management and thereby development. Since the collapse of the Soviet Union, India has been pursuing the Look East policy. Consistently rising trade integration with broader integration measures would only consolidate the existing links and compound the potential benefits for all countries involved. In the last few decades, India has successfully developed its software and services sector. Along with East Asian specialization in manufactures, India's strength in services could result in a formidable strategic combination. With the realization of such potential, India is increasingly linked with existing East Asian production networks. Further, excellent trade and transport links with Southeast Asian economies can act as a link to provide access to markets in East Asia (Anand 2009).

During the past few years a partnership between India and the Association of South East Asian Nations (ASEAN), comprising Brunei, Cambodia, Indonesia, Laos, Malaysia, Myanmar, the Philippines, Singapore, Thailand and Vietnam, has been developing at a fast pace. Though India had long cultural and social relations as well as trade links with the East Asian countries like Indonesia, Myanmar (formerly Burma), Malaysia, Thailand, Vietnam and even China in the pre-colonial period, the major part of the post-colonial trade happens to be with the Western industrially developed nations. India virtually became the sole supplier of raw materials to the industries and heavily imported the technology, crude oil and many technologically sophisticated items, including finished electronic goods, even at adverse trade terms fixed by the colonial powers. Eventually, over the years trade deficits continued to increase. However, after independence some restrictions were placed on imports to protect and encourage some domestic industries. But, India had to import technologically advanced items and know-how, as well as crude oil, gold etc., to meet the domestic requirements. The dominant trade relations with the advanced European and North American countries continued until today and India's share of world exports to these countries is far less compared to its share of imports. Again, the share of the East Asian nations in India's total external trade is rather small compared to that of the West.

In the last two decades the policies of liberalisation, privatisation and globalisation were pursued by several countries to allow the private companies' involvement in various social and economic activities. This is mostly observed for accelerating industrial development and the inflow of foreign capital for the purpose of economic growth. Also, the two big nations of Asia, India and China, achieved very high rates of growth along with other emerging Asian tigers like Indonesia, Malaysia, South Korea and Singapore, together with even a small economy like Taiwan.

We also observed the financial crisis of those countries that started with the fall of Thai Baht in 1997 followed by a similar fall in the currencies of Indonesia, Malaysia, Philippines and South Korea. This led to the flight of foreign capital, investment and thus decelerated the process of industrial and overall economic growth.

2. Objective

In this chapter we have tried to examine how the association of India with the Asian countries has grown in terms of their openness with regards to economic, social and political integration. Also, we tried to find out the extent of economic growth and positive trade impacts on India to assess whether the Look East policy has really brought about changes in the economy.

3. Materials and Methods

The KOF[1] index of globalisation (Dreher 2006; Dreher et al. 2008) in respect of economic, social and political aspects across the countries in Asia during the period 1970 to 2007, has been utilised for the purpose of analysis. The information on GDP across the countries for the years 1970, 1992 and 2008 were collected from various issues of *World Development Reports*. Also, human development index figures were collected from various issues of the *Human Development Report* published by United Nations. From the collected data we derived the (hierarchical cluster) dendogram of overall globalisation, economic globalisation, social globalisation and political globalisation of the major Asian countries for the pre (1970-1990) and post (1991–2007) globalisation periods. This provides the clustering of countries in terms of the homogeneity of globalisation. A comparison of clustering of these two periods would provide the prospect of India's increasing ties with its eastern and south-eastern neighbours. Also, the rate of growth of GDP of these countries, HDI and India's export and import trade have been compared to check how far India has succeeded in carrying globalisation forward along with economic growth.

4. Observation and Discussion

The clustering of Asian countries in terms of overall globalisation indicators during 1970 to 1990 show that India was in the close group of emerging nations like South Korea, Indonesia, Thailand, Vietnam and Cambodia together with Sri Lanka, Iran, Syria, Maldives and Pakistan. Also, it had some moderate connection with Bangladesh, China, Nepal and Myanmar. On the other hand, Malaysia, Oman and Israel belong to another cluster with the Philippines, Japan and Turkey (see Fig. 8.1a). However, during 1991 to 2007 India remained in the group of Sri Lanka, Pakistan and Syria as very close neighbours with a moderately open policy. Also, it had moderately close relations with Vietnam, Cambodia, Maldives, Iran, Nepal and Bangladesh. There has also been a move

[1] The term KOF is drawn after the name of a Swiss Company, KOF Swiss Economic Institute (Konjunktur-forschungsstelle).

towards building commonalities in trade, social and other cooperative policy with its eastern neighbours together with the West Asian countries like Iran and Syria (see Fig. 8.1b).

Trade relations along with cultural and economic cooperation among these Asian countries have been found to be highly dependent upon their geographical neighbourhood and economic compulsion (oil imports), for which India and Iran are in the same cluster. The new Asian giants like Indonesia, Philippines, South Korea, Thailand, China, Malaysia, along with Japan and Israel, became close neighbours in terms of overall globalisation during this period. Despite structural transformation of the economic policy and opening up of the eastern front, India could not come much closer to some major eastern nations like Indonesia, Philippines, South Korea, Thailand, Malaysia and even close neighbour Myanmar in terms of growth of economic and social exchange and progress. Singapore has primarily grown as the tourism and foreign trade centre of this eastern block, and has no match with any other cluster. Similarly, the military regime in Myanmar stayed separate from even its neighbouring countries in many respects.

The economic globalisation that spurred the progress in the event of vibrant international trade in terms of rise in the percentage of the country's GDP, flow of goods and services and capital, and reduction in barrier of trade brought many countries closer together. In this respect, India, Nepal, Bangladesh, Sri Lanka, Pakistan, China and Iran were close neighbours during 1970–90, while the fast growing Asian giants like Indonesia, South Korea and the Philippines formed a cluster with Thailand, Japan, Cambodia, Vietnam, Turkey and Syria that followed similar open economic policies in front of international trade and flow of foreign capital. Malaysia, however, was never in that cluster because, despite having a similar growth pattern in the 1990s, it suffered a financial crisis and recession after 1997 along with Indonesia, Thailand, and Philippines (see Fig. 8.2a).

Even though the waves of globalisation rocked the Indian economy, India could not enter the cluster of Asian giants in recent years, despite having followed the path of Look East to achieving closer ties with its eastern neighbours and other South-East Asian countries. Even countries like Sri Lanka and China entered the cluster of Japan and Syria along with those fast growing Asian nations at the next level (see Fig. 8.2b).

Fig. 8.1a. Clustering of Asian countries using average group links of overall globalisation 1970–1990

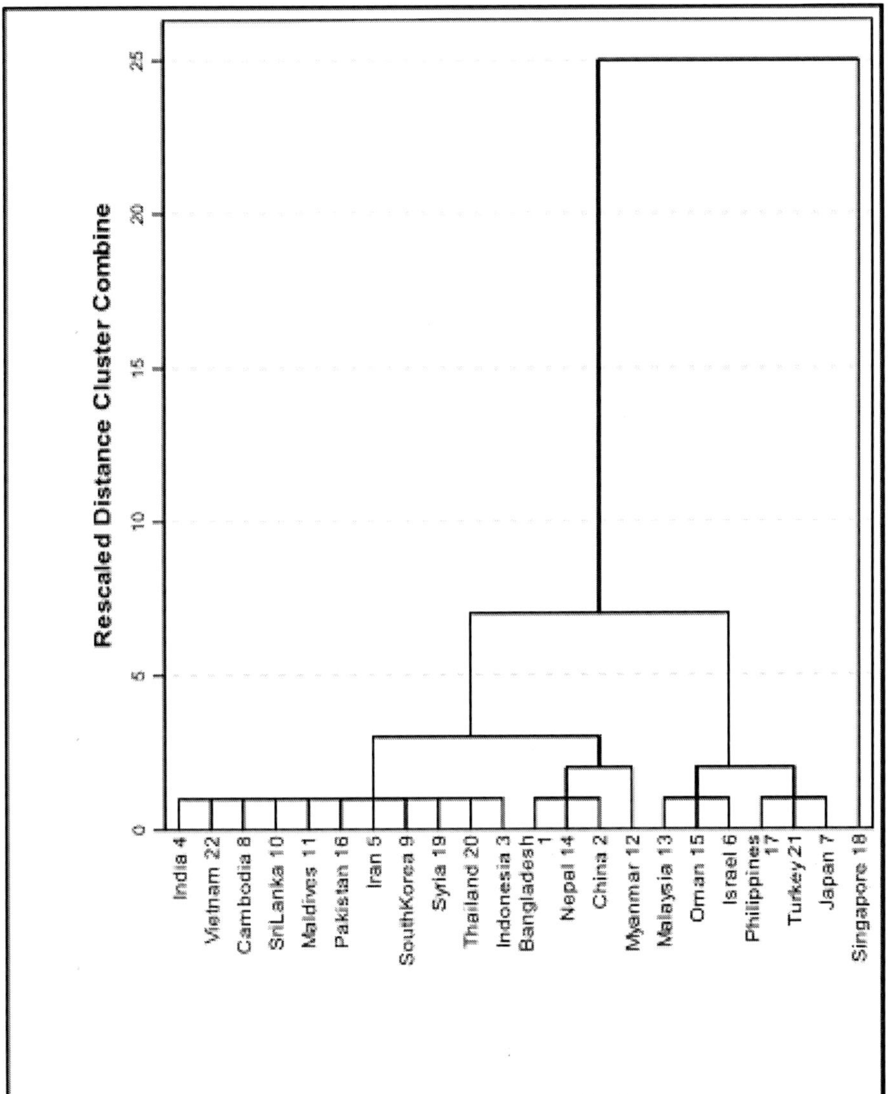

Fig. 8.1b. Clustering of Asian countries using average group links of overall globalisation 1991–2007

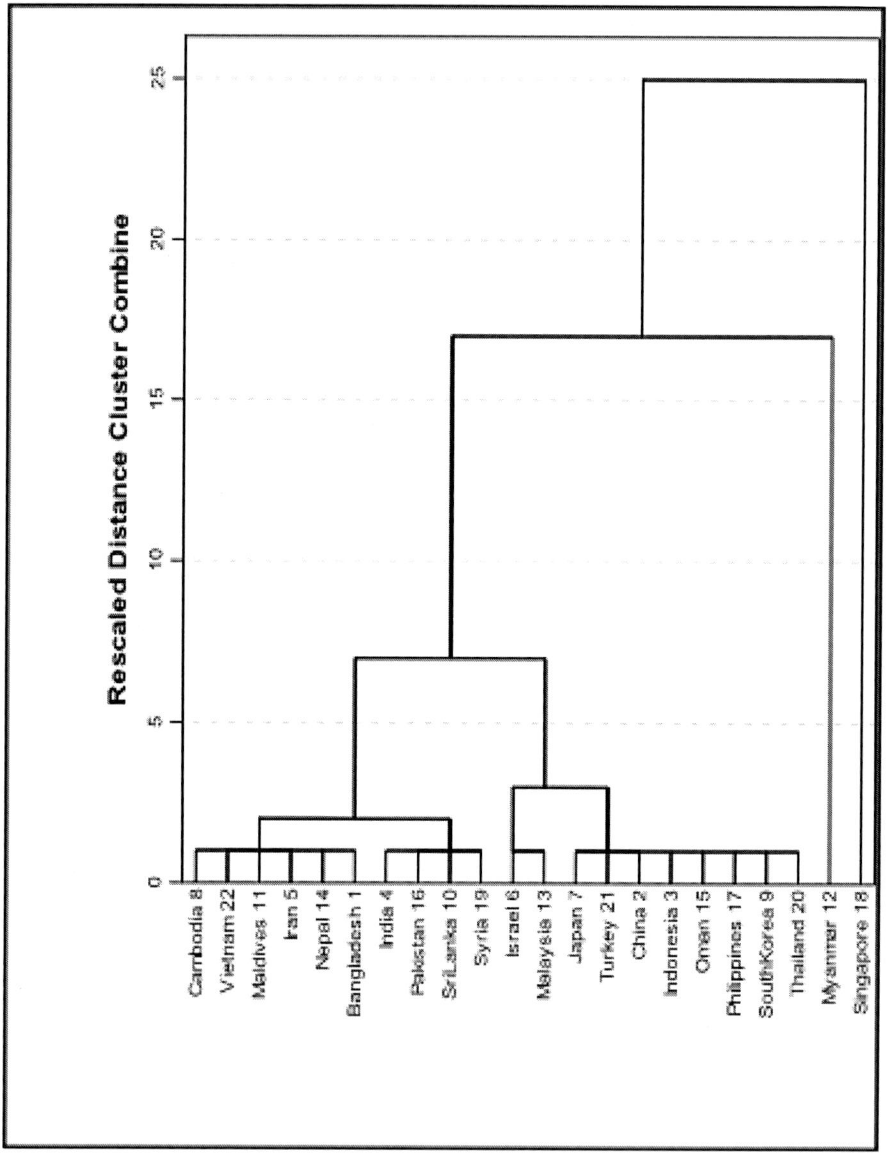

Fig. 8.2b. Clustering of Asian countries using average group links of economic globalisation 1991–2007

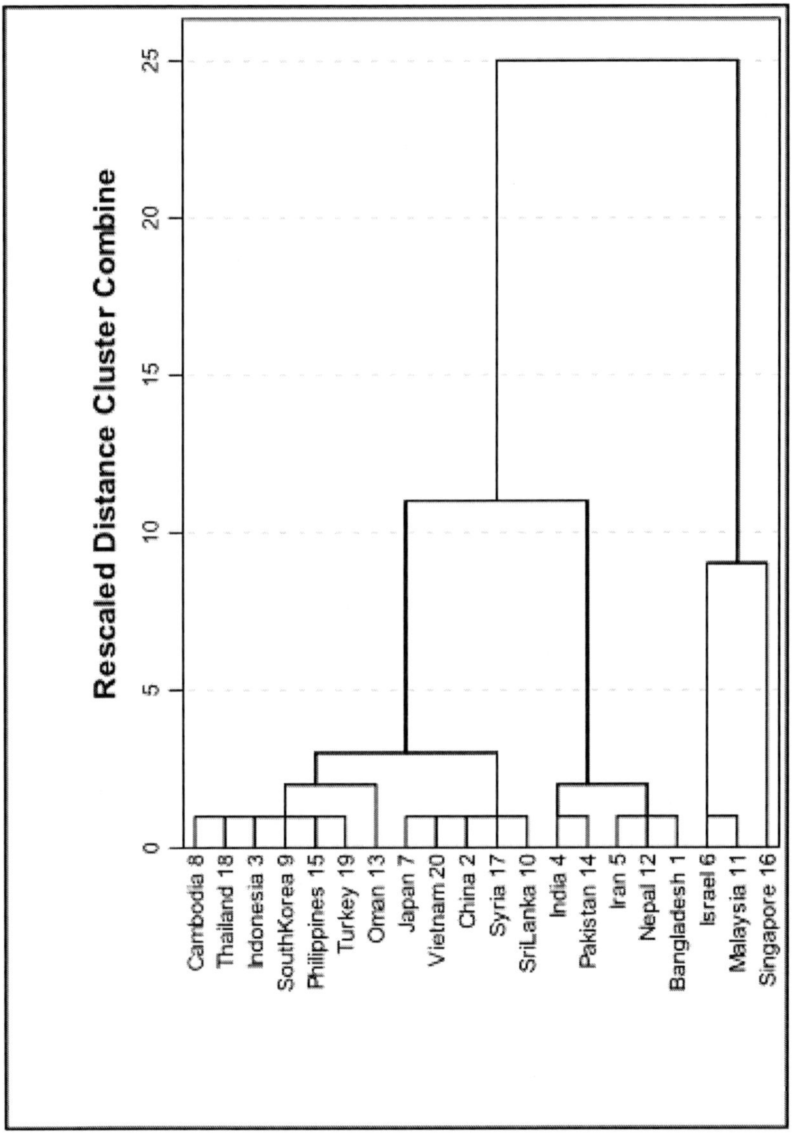

Fig. 8.3a. Clustering of Asian countries using average group links of social globalisation 1970–1990

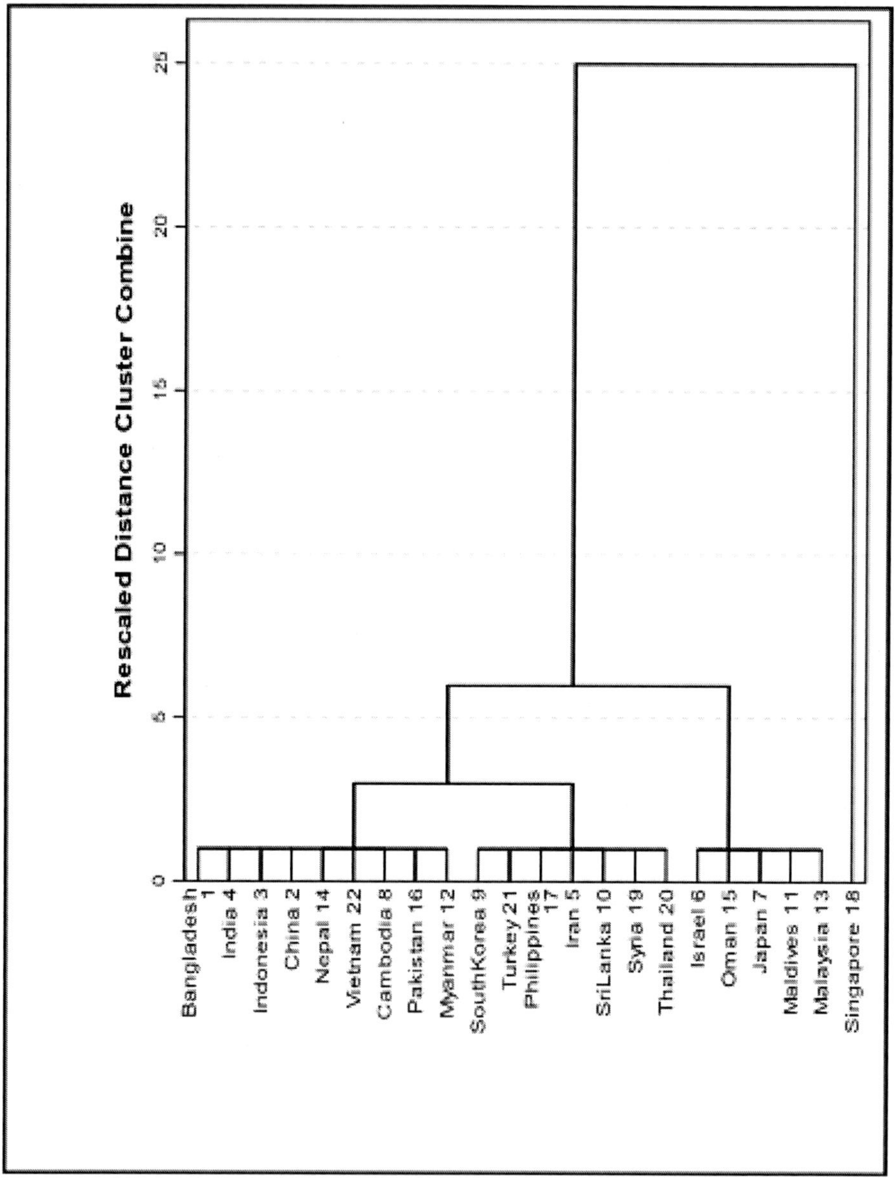

Fig. 8.3b. Clustering of Asian countries using average group links of social globalisation 1991–2007

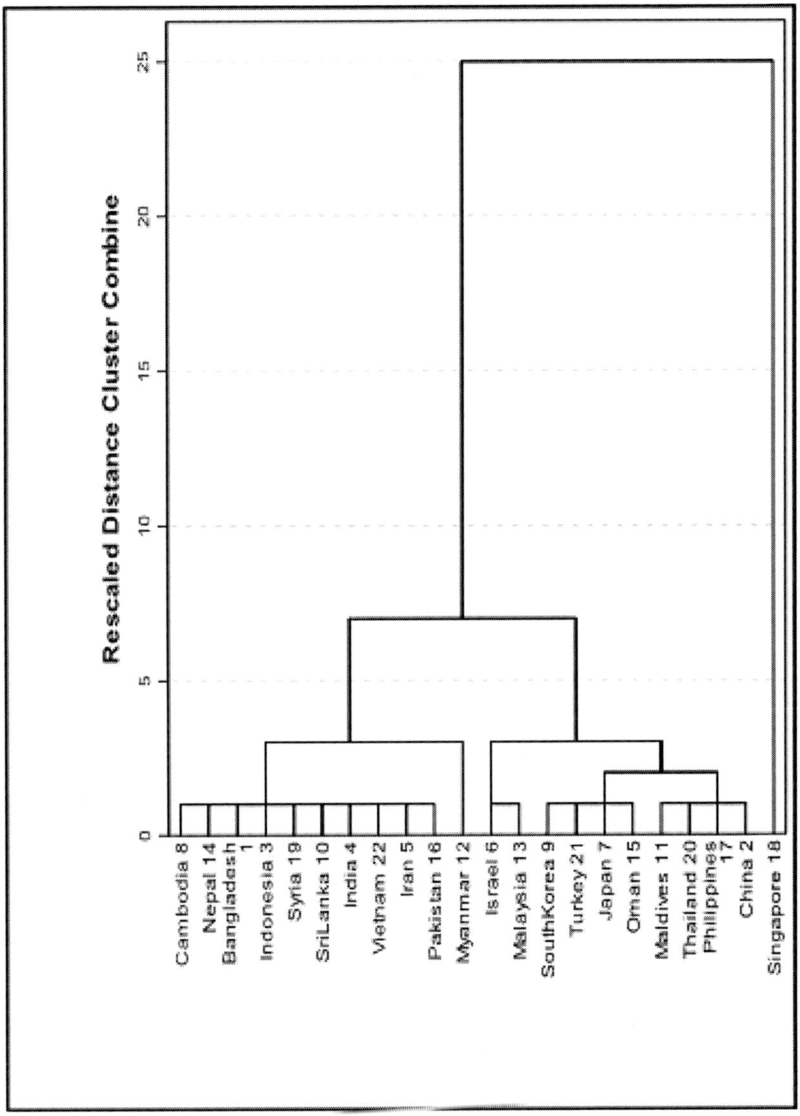

Like overall globalisation, during 1970 to 1990 India was in the close group of emerging nations like Indonesia, Vietnam, Cambodia, Bangladesh and Pakistan in terms of social globalisation (see Fig. 8.3a above);it was also closer to China and Myanmar. The geographical proximity played an important role in this social opening and interactions. On the other hand, Sri Lanka, Iran, Syria and Maldives were closely tied to South Korea, the Philippines and Thailand. After 1990, India could extend its social proximity to Iran while the other eastern neighbours—as well as China, a formerly closed neighbour—remained outside its cluster in terms of social openness. South Korea, the Philippines, Thailand, Japan and China, like their economic openness, also followed more or less similar policies and thus formed a cluster in terms of a dendogram. So, in the post globalisation era there were two overall blocks among the Asian countries in terms of social gobalisation. The middle or better developed nations were in one cluster and the less-developed economies with similar socio-cultural set-ups and traditional bondage fell in another (see Fig. 8.3b).

Fig. 8.4a. Clustering of Asian countries using average group links of political globalisation 1970–1990

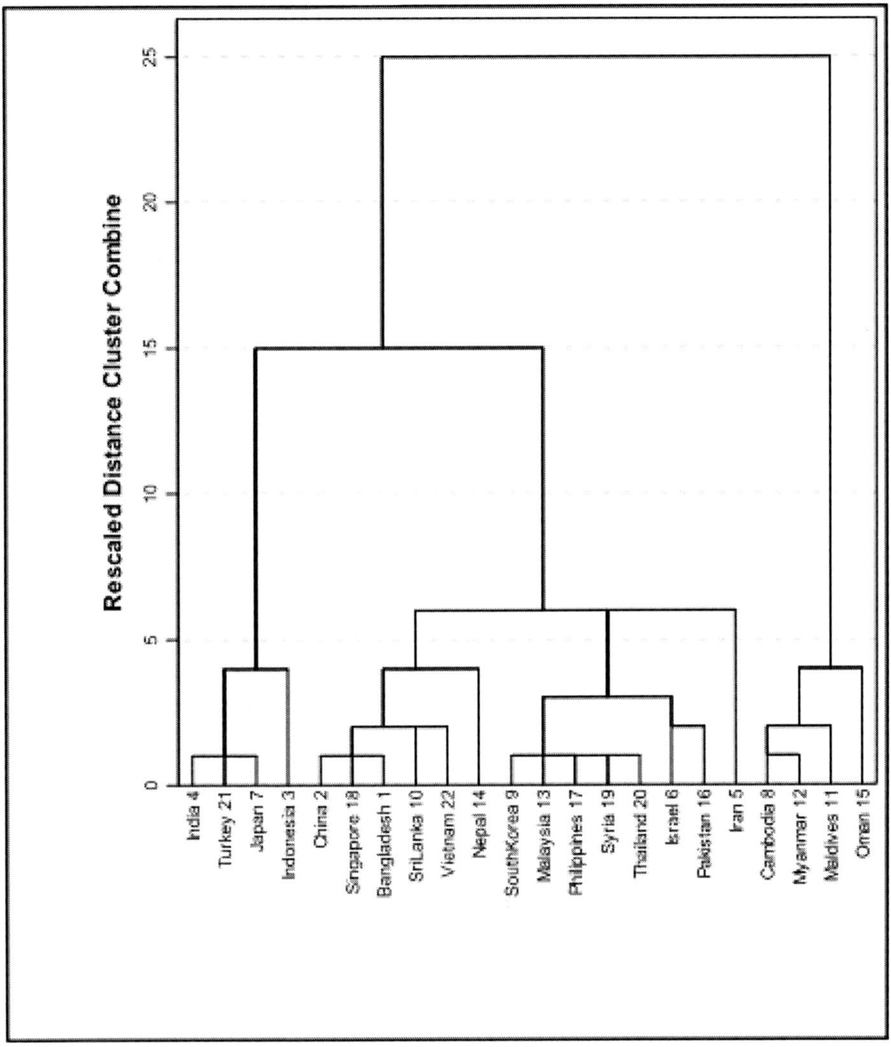

In terms of political globalisation, India was outside the cluster of all the geographical neighbours like Pakistan, Bangladesh, China, Nepal, Sri Lanka etc. due to their political disturbances and conflicts. It had close political openness with Indonesia, Japan and Turkey only during 1970 to 1990 (see Fig. 8.4a above). Countries like China, Bangladesh, Sri Lanka and Vietnam, who were never in a cluster with Singapore in terms of their economic or social globalisation policies, formed a politically close cluster. The other Asian giants with greater progress in economic and social globalisation as well as development also followed a faster political globalisaion and formed a cluster along with Syria and Israel before 1990 (see Fig. 8.4a above). During the post-globalisation period the situation did not change much for India because of its strange political relationship with the geographical neighbours, except for a little improvement with Pakistan (see Fig. 8.4b below).

Fig. 8.4b. Clustering of Asian countries using average group links of political globalisation 1991–2007

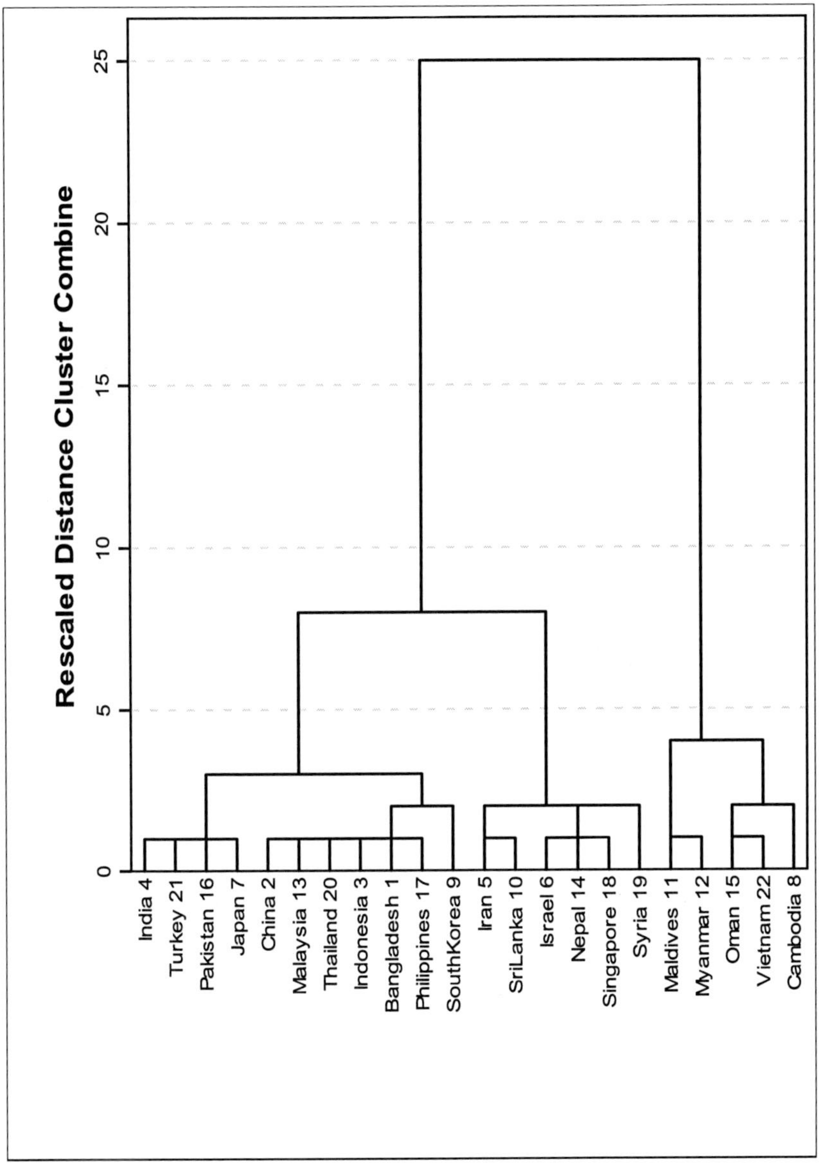

Progress of India's External Trade with other ASEAN Nations

The available information shows that India's trade with the ASEAN countries increased significantly from 1996 to 2011. Exporting to these countries increased by an annual compound rate of 17.66% and 21.89% respectively during 1996–97 to 2004–05 and 2004–05 to 2010–11, while India's overall exports increased by an annual compound rate of 15.46% and 20.39% during those two periods. It indicates that despite the financial crisis and economic slowdown of the major ASEAN countries, Indian exports to these countries rose at an increasing rate, especially after it became a fully-fledged member of the ASEAN in 2002. However, the rise in overall exports was also at an increasing rate mainly due to the outsourcing of many external activities to various software and other organisations operating in India. The growth of exports to the ASEAN nations has been marginally higher than its overall export (see Table 8.1 below).

India's imports from the ASEAN countries has also increased at an increasing rate since 1996–97. The imports increased by an annual compound rate of 18.67% and 22.65% during 1996–97 to 2004–05 and 2004–05 to 2010–11. Also, its overall import increased by an annual compound rate of 17.39% and 22.38% during those two periods. Here also, the growth of imports from the ASEAN nations has always been marginally higher than India's overall imports. Since 1996–97 the gap in the growth of India's imports and exports from the ASEAN countries has always been about 1%, and about 2% overall. Thus, it is apparent that India's Look East policy has not so far been able to reduce the rate of growth of imports from these countries below the rate of growth of exports, though India's main objective had been to expand its exports to the markets of the Eastern countries.

Table 8.1. India's Exports and Imports with the ASEAN Countries since 1997 (Rs. in Crore)

	Export			Annual compound Rate of Growth (%)		Import			Annual compound Rate of Growth (%)	
	1996–97	2004–05	2010–11	1996–97 to 2004–05	2004–05 to 2010–11	1996–97	2004–05	2010–11	1996–97 to 2004–05	2004–05 to 2010–11
ASEAN	10303.65 (8.67%)	37858.78 (10.09%)	124129.08 (10.86%)	17.66	21.89	10415.66 (7.50%)	40953.53 (8.17%)	139439.33 (8.28%)	18.67	22.65
Total	118817.97	375339.53	1142648.97	15.46	20.39	138919.66	501064.54	1683466.96	17.39	22.38

Source: Export-Import Data Bank, Ministry of Commerce, Government of India.

Table 8.2. India's Balance Of Trade with the ASEAN Countries, 1995–2008 (Million USD)

Country	1995	2002	2005	2007	2008
Singapore	45.5	46.7	2268.1	-511.5	549.1
Indonesia	200.3	-493.6	-1628.9	-2962.2	-3772.0
Thailand	302.2	376.1	-137.3	-519.0	-659.5
Malaysia	-504.9	-587.7	-1292.2	-3875.3	-4427.0
Vietnam	108.5	276.4	506.1	1088.3	1441.0
ASEAN-5	151.7	-382.2	-284.2	-6779.7	-6868.4
Philippines	122.4	334.5	278.9	397.9	527.4
Myanmar	-131.9	-278.9	-371.9	-646.3	-668.9
Cambodia	1.8	16.3	20.9	43.6	49.6
Brunei	7.2	4.3	3.6	-225.3	-308.7
Laos	NA	NA	NA	NA	NA
Other ASEAN	-0.6	76.2	-68.5	-430.1	-400.7

Source: Based on UN Comtrade Database (Francis 2011)

Moreover, the economy of north-east India through the expansion of trade through the borders of these states was expected to improve, but there has hardly been any competitive manufacturing output that could capture the export markets of these countries. In fact, several relaxations and the promotion of a free trade agreement accelerated the import of cheap consumable items especially from Malaysia, Taiwan, Thailand, Myanmar, South Korea and even China.

Table 8.2 above reveals that among the ASEAN countries, India maintained a trade deficit with the fastest growing nations like Indonesia, Malaysia, Myanmar and even Thailand. India had some trade surplus with Vietnam, Cambodia and the Philippines in some years, as well as with Singapore. Overall, India has a growing trade deficit with the ASEAN countries after following the Look East policy. Along with that, the reduction in tariff to zero by India on a number of items (including some agricultural products where the domestic producers may face uncertain price competition) imported from these countries would increase the trade deficit further and replace many domestic items purchased by the consumers. Viswanathan & Shah (2008) demonstrated the adverse impact of trade liberalisation on India's plantation sector, especially tea and rubber production.

Recent trends of India's exports to and imports from the ASEAN *vis a vis* its total exports and imports show that since 2007, the imports often ASEAN countries from India have increased at an annual average rate of 22.63%, while India's imports from these countries have increased by an average annual rate of 11.38%. On the other hand, India's total exports and imports to and from all the countries in the world have increased by annual average rates of 19.65% and 20.44% respectively after 2007. Moreover, the percentage share of these ASEAN countries together to India's total exports has increased marginally from 9.98% in 2007 to 10.86% in 2011 (see Table 8.3 below). The share of the ASEAN to India's total imports has however declined marginally from 9.75% in 2007 to 8.28% in 2011 (see Table 8.4 below). A more interesting point is that India's imports from its neighbours like Myanmar, Laos, Brunei and Vietnam have declined during the last few years, though exports have registered an increase. Still, there had been a trade deficit of a total 1,531,025 USD with these ASEAN countries in 2011, which was 2,484,230 USD in 2007.

Table 8.3. India's exports to the ASEAN countries and its share since 2007 (USD)

Countries	2007	2008	2009	2010	2011	Annual Avg. Growth Rate 2007–11 (%)
Singapore	2,746,160.82	2,966,223.24	3,775,688.18	3,594,829.70	4,690,605.33	15.25
Thailand	653,562.38	727,877.20	872,399.50	822,762.26	1,268,210.59	19.92
Malaysia	590,192.56	1,033,728.54	1,578,036.20	1,350,392.32	1,806,659.10	36.79
Indonesia	917,696.77	869,277.93	1,157,782.95	1,460,463.91	2,838,765.89	37.11
Viet Nam	444,623.84	645,128.09	794,947.72	867,397.61	1,208,229.28	29.18
Philippines	263,596.76	249,073.46	337,935.11	354,650.05	401,240.74	12.06
Brunei Darussalam	3,759.13	4,206.74	7,994.52	11,605.02	11,529.57	36.62
Cambodia	23,603.66	21,510.37	21,507.71	21,553.19	29,118.31	6.61
Myanmar	633,74.59	746,19.38	101,776.52	98,472.88	152,175.87	26.36
Lao PDR	1,076.38	1,542.58	4,445.31	8,120.26	6,373.47	73.16
ASEAN-10 Aggregate	5,707,647 (9.98%)	6,593,188 (10.05%)	8,652,514 (10.29%)	8,590,247 (10.16%)	12,412,908 (10.86%)	22.63
India's Total Export	57,177,928.52	65,586,352.18	84,075,505.87	84,553,364.38	114,264,897.18	19.65

Source: Department of Commerce, Ministry of Commerce and Industry, Govt. of India.

Table 8.4. India's imports from the ASEAN countries and its share since 2007 (USD)

Countries	2007	2008	2009	2010	2011	Annual Average Growth Rate 2007–11 (%)
Singapore	2,483,996.69	3,268,217.81	3,456,141.62	3,062,330.81	3,254,576.75	31.57
Thailand	789,880.06	926,400.41	1,235,265.25	1,388,850.94	1,945,991.14	17.28
Malaysia	2,395,876.10	2,417,613.44	3,259,156.48	2,449,403.25	2,974,590.44	0.91
Indonesia	1,886,485.99	1,942,053.15	3,075,129.40	4,100,880.75	4,513,629.30	2.95
Viet Nam	75,861.20	69,807.09	186,226.30	245,911.66	484,759.06	-7.98
Philippines	75,737.13	82,387.82	116,542.91	149,329.11	195,046.60	8.78
Brunei Darussalam	129,068.31	90,868.70	188,194.37	202,852.73	106,485.70	-29.60
Cambodia	714.94	1,155.25	1,197.33	2,405.90	3,637.69	61.59
Myanmar	354,094.52	325,928.14	424,076.82	610,794.46	465,115.40	-7.95
Lao PDR	162.52	45.69	214.92	9,218.20	100.53	-71.89
ASEAN-10 Aggregate	8,191,877 (9.75%)	9,124,478 (9.01%)	11,942,145 (8.69%)	12,221,978 (8.96%)	13,943,933 (8.28%)	11.38
Total Import	84,050,631.33	101,231,169.93	137,443,555.45	136,373,554.76	168,346,695.57	20.44

Source: Department of Commerce, Ministry of Commerce and Industry, Govt. of India.

Growth of Gross Domestic Product (GDP) and Human Development Index (HDI)

The opening of the economy along with the Look East emphasis, i.e. regional or sub-regional cooperation within the ambit of globalisation, have been undertaken with the objective of capturing the world market, expanding exports, acquiring technology, improving human resources, achieving overall economic growth and creating employment opportunities. Also, the target has been to attract foreign capital investment, especially in the manufacturing sector, to boost the economy. Table 8.5 below shows that the direct foreign investment coming to India has increased significantly from merely 0.075 billion dollars to 34.613 billion dollars during 1991 to 2009, following several bilateral and multilateral agreements in the wake of globalisation and the recent Look East policy. Also, the FDI of India in other countries increased from 0.011 billion dollars to 14.897 billion dollars during this period, which is an indication of India's improved capability to enter the markets of other countries not only through rising exports but also through the expansion of industrial and service sector activities. This was undertaken by our multinational companies in other countries depending upon their competitive strength (see Table 8.5 below). However, India is still lagging behind the other fast-growing Asian countries like China in terms of attracting foreign capital or direct investment and registering progress in technologically advanced manufacturing activities and their exports.

Despite foreign capital inflow in various sectors during the last three decades, the resultant growth has been mixed. Empirical studies by Sharma (2000) and Alfaro (2003) suggest no significant impact on India's GDP during the 1980s and 1990s, while Chakraborty & Basu (2002) found some positive impacts on the GDP in the short term. Though several countries followed the open economic policy with a number of relaxations, the growth of those economies has varied significantly. The annual compound rate of growth of the GDP of India was significantly lower than most of those ASEAN nations in the pre-globalisation period during 1970–1992, while it exceeded the GDP growth of those countries during the strong globalisation phase 1992–2008. One point which is clear is that India could take some advantage of globalisation in promoting employment, exports, improvement in technology and even growth of GDP, but it failed to achieve much progress in human development. Its trade balance with the major ASEAN nations is still negative and the production as well as exporting of manufacturing products is still far lower than the Asian Giants. Whatever export growth we observed was due to the growth of the service sector, and the industrial and overall growth of north-east India as a result of the Look East policy is yet to bear fruit.

Table 8.5. FDI inflows and outflows of India 1991–2009 (billion USD)

	Inflows		Outflows	
Year	World	India	World	India
1991	154.0	0.075	198.1	0.011
1992	166.0	0.252	202.7	0.024
1993	223.5	0.532	242.6	0
1994	256.1	0.074	286.9	0.082
1995	342.5	2.151	362.6	0.119
1996	389.0	2.525	396.5	0.240
1997	486.5	3.619	476.1	0.113
1998	707.2	2.633	682.3	0.047
1999	1,087.5	2.168	1,076.8	0.080
2000	1,401.5	3.588	1,232.9	0.514
2001	825.3	5.478	753.1	1.397
2002	628.1	5.620	537.1	1.678
2003	565.7	4.321	565.7	1.876
2004	732.4	5.778	920.3	2.175
2005	985.8	7.622	893.1	2.985
2006	1,459.1	20.328	1,410.6	14.282
2007	2,100.0	25.001	2,267.5	17.233
2008	1,770.9	40.418	1.928.8	18.499
2009	1,114.2	34.613	1,101.0	14.897

Source: UNCTAD, FDI/TNC Database (www.unctad.org)

Table 8.6. Growth of India's GDP and HDI *vis a vis* the ASEAN and neighbouring countries before and after globalisation wave (percentage)

Country	GDP		HDI	
	1970–1992	1992–2008	1975–90	1990–2005
Singapore	15.60	8.97	0.84	0.73
Thailand	13.29	5.52	0.98	0.62
Malaysia	12.64	7.92	1.06	0.75
Indonesia	12.40	9.17	1.91	1.01
Viet Nam	--	--	--	1.12
Philippines	9.81	7.50	0.64	0.45
Brunei Darussalam	NA	NA	NA	NA
Cambodia	--	--	--	--
Myanmar	13.90	NA	NA	NA
Lao PDR	NA	9.92	NA	1.54
India	6.57	11.46	1.46	1.16
Bangladesh	5.95	7.79	1.31	1.74
Sri Lanka	6.45	10.07	0.84	0.38
Nepal	5.44	9.96	2.36	1.50
Pakistan	7.19	9.08	1.62	1.11
China	8.00	8.80	1.20	1.37
South Korea	17.28	7.41	0.98	0.74

Source: Calculated on the basis of data available from various issues of World Development Report, Human Development Report.

5. Concluding Remarks

In conclusion, it can be said that even after following cautious globalisation measures and forming several alliances at the regional and sub-regional levels, and formation of the free trade zone especially with the Eastern nations via the Look East policy, India could not make significant progress comparable to the Asian giants. Despite this, the fact that there was a notable crisis in these fast growing Asian countries after 1997 and their recovery afterwards, India's effort to catch up with the Eastern markets has not been highly successful. Its exports could not rise significantly *vis-a-vis* those of other countries and the trade deficit with the major ASEAN countries still exists. Also on the growth front, India lagged behind those countries in terms of GDP and Human Development growth during the 1980s and 1990s. However, in the last decade we find some improvements and India's GDP and HDI grew at relatively faster rates than those nations.

Finally, it can be said that India needs to improve its manufacturing base through the adoption of the best possible technological updating, especially in the north-eastern states, for which the Look East policy was also undertaken. If the production base is not improved in these states this region will remain a supplier of only raw materials, and not the finished product. The region will continue to face competition and thus be flooded with cheap products imported from its eastern neighbours.

References

Alfaro, L. "Foreign Direct Investment and Growth: Does the Sector Matter?" *Harvard Business School Working Paper*, Harvard, USA. 2003

Amavilah, V. H. "National Identity, Globalization, and the Well-being of Nations." MPRA paper No.14948, 2009a. http://mpra.ub.unimuenchen.de/14948/1/MPRA_paper_14948.pdf.

—. "National Symbols, Globalization, and the Well-being of Nations." 2009b. http://mpra.ub.uni-muenchen.de/14882/1/MPRA_paper_14882.pdf.

Anand, M. "India-ASEAN Relations: Analysing Regional implications." IPCS Special Reports 72, Institute of Peace and Conflict Studies, New Delhi, 2009

Antweiler, W., B. R. Copeland, & M. S. Taylor. "Is Free Trade Good for the Environment?" *American Economic Review* 91 (4) (2001): 877–908.

Beams, Nick. "Globalisation: The Socialist Perspective." 2000. http://www.wsws.org/articles/2000/jun2000/lec1-j05.shtml (accessed December 12, 2010).

Chakraborty, C. & P. Basu. "Foreign Direct Investment and Growth in India: a Cointegration Approach." *Applied Economics* 34 (2002): 1061–1073.

Cherni J. A. "Globalisation and Environmental Sustainability in Cities of Developed and Developing Countries." *Revista Theomai (edicion electronic), numero 4,* Universidad Nacional de Quilmes, Argentina, 2001.

Cole, M. A. "Trade, the Pollution Haven Hypothesis and the Environmental Kuznets Curve: Examining the linkages." *Ecological Economics* 48 (2004): 71–81.

De, U. K. & M. Pal. "Dimensions of Globalization and their Effects on Economic Growth and Human Development Index," *Asian Economic and Financial Review* 1 (1) (2011): 1–13.

Dreher, A. "Does Globalization Affect Growth? Evidence from a New Index of Globalization." *Applied Economics* 38 (1) (2006): 1091–1110.

Effland, A., M. A. Normile, D. Roberts & J. Wainio. "World Trade Organization and Globalization Help Facilitate Growth in Agricultural Trade." 2006. http://www.ers.usda.gov/AmberWaves/June08/Features/WTO.htm (accessed December 12, 2010).

Eweje, G. "Globalisation & Anti-Globalisation Movements." *Topics in International Business, Lecture 2.* Albany, Massey University, 2005.

Francis, S. "The ASEAN-India Free Trade Agreement: A Sectoral Impact Analysis of increased trade integration in goods." *Economic and Political Weekly* 46 (2) (2011).

Heintz, James. "Globalization, Economic Policy and Employment: Poverty and Gender Implications." Employment Strategy Papers, Geneva International Labour Office, 2006.

Kulkarni, Kishore G. "Effect of Globalization on India's Economic Growth." Presented in the Oxford Roundtable Conference held in Oxford University, UK, in July 2005.

Sharma, K. "Export Growth In India: Has FDI Played A Role?" *Discussion Paper No. 816, Economic Growth Center, Yale University,* 2000.

Tang, R. "Is the Common Good Improved by Economic Globalisation and the Activities of Multinational Corporations?" *International Journal of Business and Management* 3 (1) (2008): 141–145.

Versi, A, "The Human Face of Globalisation," Business Source Premier, ISSN:01413929, Massey University, 2004

Viswanathan, P. K. & A. Shah "Trade Reforms and Crisis in India's Plantation Agriculture: Case studies of tea and rubber plantation sectors." Paper presented at the Fourth Annual South Asia Conference on Trade and Development 2008 organised by CENTAD, New Delhi, December 17–18, 2008.

World Bank & IMF. "Aid for Trade: Harnessing Globalization for Economic Development." Policy paper, 2007.

CHAPTER NINE

HEALTH PROVIDERS AND THEIR EVALUATION: A STUDY OF BURDWAN MEDICAL COLLEGE HOSPITAL

ATANU SENGUPTA
AND DEBJYOTY MUKHERJEE

1. Introduction

> *The doctor arrived towards dinnertime and said, of course, that although recurring phenomena might well elicit apprehension, nonetheless there was, strictly speaking, no positive indication, yet since neither was there any contraindication, it might, on the one hand, be supposed, but on the other hand it might also be supposed. And it was therefore necessary to stay in bed, and although I don't like prescribing, nevertheless take this and stay in bed.*
>
> —*The Devil*, Leo Tolstoy

Estimating efficiency in the medical care industry is extremely important for the basic human development of a country. However, the process is fraught with a greater number of pitfalls and contradictions than is commonly recognized. Medical staff (doctors, nurses, technicians etc.) cannot work in isolation. Their efficiency is largely dependent on a number of subjective and objective factors[1] that we collectively define as the "service state." Nonetheless, in any given "service state," the performances of medical staff differ; however, it becomes difficult to

[1] At any given point of time certain aspects of these combinations are measurable, like number of beds, availability of a specific diagnostic instrument and trained manpower to run it *vis-à-vis* the subjective perception of various aspects of the people who are attached to the service provisions like adequacy of infrastructure to provide successful care or cleanliness of the system etc.

segregate these systemic factors from the individual performances. In this chapter we have analysed the efficiency of the unit source of outputs within a given service state.

There are many studies that have attempted to measure the system efficiency. For illustration we considered the World Health Report (2010), Background Paper 28, written by Dan Chisholm and David B. Evans. This paper argues that increase in the expenditure on health per capita does not match with the corresponding improvement in life expectancy. Many cases are cited in its favour. An example from our region is the comparison between China and Egypt—both the countries spent the same on health per capita but each newborn in China can expect to live an extra five years. This is same as the life expectancy in Peru, Turkey and Hungary, but at a considerably lower cost. It is therefore easy to see that the expenditure is not incurred efficiently, and to tackle this problem the measurement of efficiency is a must.

This work is subject to a considerable dilemma. It clearly points out the possible reasons for efficiency according to its sources. Many of the reasons are subjective in nature. For example, one of the possible reasons for efficiency with respect to medicine is substandard or counterfeit drugs. The reasons are weak drug regulatory structure and weak procurement mechanism. The way out is improvement of drug regulation and quality control. Anyone can easily find that these factors are subjective in nature, anda similar subject reasoning has also been found in the case of other resources.

However, while indulging in the efficiency estimation, the paper uses the traditional measures of technical and allocative efficiency (such as the nonparametric envelopment or parametric frontier techniques). This dilemma in the subjective concept of efficiency and its objective estimation has been the hallmark of almost of all studies in health efficiency. This chapter tries to reduce the dilemma through the subjectivism of the efficiency measures.

Thus, standard efficiency analysis is an inter-firm comparison and pinpointing the sources of productivity growth. For example, it may infer whether hospital A is more or less efficient than hospital B (or School A is better than School B). However, it cannot, in general, tell us about the relative performances of physicians in the same hospital (or teachers within the same school).

This concept of efficiency is thus in contradiction to the common perception of efficiency that is more subjective in nature. In real life, people not only compare with systems but also with parts of the same system. This is particularly so in the service sector where both level and delivery are important. For the medical care sector, Arrow (1963) defines this as "mutual trust" between the patients and doctors. Broadly speaking, the paper emphasizes the individual effects of efficiency.

This chapter is divided into sub-sections concerning subjective evaluation, objective, methodology, data sources, findings and conclusions.

2. Towards a subjective evaluation

The focus of this section is the development of a health model where the physicians are responsible for diagnosing illness, deciding on an appropriate treatment and assuring that treatment is carried out as prescribed. The physician is viewed as knowledgeable and powerful whereas the patient is passive, accepting the care, compliant and dependent on the physician and their goodwill. The difference between a physician and patient is the asymmetry of knowledge in the field of medical care. Given the available stock of infrastructure, both soft and hard, along with available consumables, the physician generates a service or treatment module.

Thus, with the physician's knowledge and experience, perception grows of what would best be possible and what they are achieving in their present endeavour. Thus, with the physician's perception of the hospital, its operational standard could be judged with the service state, presented through the following diagram.

Fig. 9.1. Interaction within service state

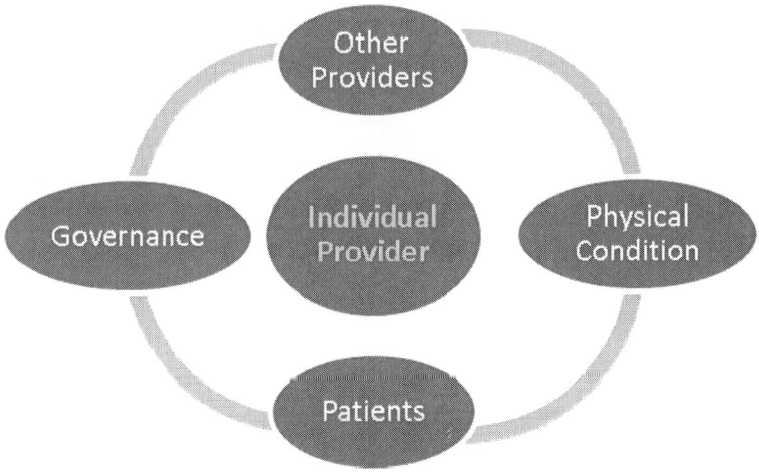

Thus, clearly:

- Efficiency perception departs from the traditional dyadic to triadic relationship
- Consideration of individual skills or knowledge (homogeneous/ heterogeneous) interacts with service state and produce output.

This chapter is a deviation from the standard analysis, in which efficiency is a mathematically precise concept. However, this preciseness is not common with its everyday uses. First is the distinction between the system and individual efficiency. Second, even systemic efficiency cannot be an objective entity only. Utility is determined by its usage. For example, provision of school buses may be understood as an improvement to a school infrastructure, but the rigidity of its timing may be a disincentive to many, particularly to the poor and working classes where parents may not hold a regular job. To them, walking or cycling may be more efficiency improving devices. Hence, the subjective evaluation will also enter the field of systemic efficiency. However, we are concerned with the subjective evaluation of the provider themselves.

The interaction between the individual and systemic efficiency can be seen in Fig. 9.2 below.

Fig.9.2. Interaction between individual and systemic efficiency

Individual efficiency		**System Efficiency**	
		Efficient	**Inefficient**
	Efficient	Case I	Case II
	Inefficient	Case III	Case IV

The diagram shows four cases:

Case I: where both the system and the individual are efficient
Case II: where the individual is efficient but the system is inefficient
Case III: where the system is efficient but the individual is inefficient
Case IV: where both the system and individual are inefficient.

The above cases provide a more comprehensive view of efficiency than the traditional analysis suggests, where efficiency is one dimensional system efficiency. Our analysis considers an additional dimension of individual efficiency. However, the relationship is not as crystal clear as the diagram would suggest, as system efficiency no longer has the object efficiency measure as used by the World Health Report (2010). Here we evaluate the system performance as visualized by the individuals. In many cases, the individual may visualize spots where the standard efficiency analysis failed to illuminate. For example, even in a well-furnished and modernized hospital, the absence of some crucial elements may prove a hindrance to the proper redresses.

This chapter considers the evaluation by the health providers themselves.

3. Objective of the Study

In this chapter we concentrate on the macro level of efficiency analysis, i.e. within the firm. The fundamental consideration is that it is the human beings that use the non-human resources and produce the service. Thus, the non-human resources as given, the human resources that produce the goods and services are in a triadic relation with the patients and the community *vis-à-vis* the process of governance.

We assume that in a hospital system, outcome is the aggregate of the individual doctors. Thus, at a micro level, the efficiency of individual doctors is a unit source for the aggregate efficiency of the hospital. We divide the universe of the doctor -provided health service into four distinct states: the doctors themselves, the service of the hospital together with the community and the health governance system. The doctor uses the service state with a triadic relation in the governing system and produces the output. If this broad relationship state is accepted then we can deduct that the efficiency level of the individual doctor depends on the service state and the governance system. In a given service state and governing system, the output of individual doctors differs. Though the skill sets of individual doctors are different they have a basic level of knowledge and training in order to be a registered doctor. Even at this basic level efficiency differs.

An important aspect of this chapter is the self-reported performance of the doctors, based on the notion that physicians are the best judges of their own capacities and produce the desired level of service.

Also, the capability or efficiency of individual doctors is not independent of the efficiency of the system. The standard definition of efficiency, i.e. the best possible output with the given inputs, is not applicable owing to the fact that supplying inputs to achieve a desired outcome is a part of the efficiency of the system (supply side efficiency of human resources).It is assumed that it is the service providers who can best judge what the efficacy of the service state is, and another aspect of the study is their perception of the operational standard of the hospital.

Precisely, the objective of the study is the service providers' perceptions about the service state and self-evaluation of their efficiency.

We measure efficiency in three different directions:

(1) Efficiency of individual
(2) Efficiency of Burdwan Medical College Hospital as evaluated by the service providers
(3) Efficiency of BMCH in comparison to other MCH as evaluated by the service providers

All these efficiency measures are a result of subjective evaluation. These contrast our work from the traditional works on health efficiency (World Health Report 2010) where efficiency is measured in an objective fashion.

4. Methodology

The service providers in a hospital setup are divided into four classes: physicians, nurses, technical support staff and general support staff. These four categories of individuals possess different set skills and knowledge to run the system, irrespective of their knowledge and skills directly or indirectly related to medical care. We first analyze their perception of the hospital set up.

Since perception is a subjective assessment, a ten-point Likert scale is used for different aspects of operational standard. Generally in such studies the Likert scale is used on a 5-point rating basis, but in our case we have used the 10-point Likert scale to gauge a higher level of variability in the responses.

To understand the efficiency of the service state, we have used seventeen different parameters (listed in Table 9.1 below) of different categories: patient related, physical conditions, other providers, governance, self-efficiency as well as a comparison with other medical college hospitals (MCHs). Only one is objective while the rest are subjective in nature. Among the subjective factors, there are both systemic and individual level parameters. In all they provide a comprehensive view of the BMCH.

Table 9.1. Description of the variables studied

Sl. No.	Parameters	Classification	Nature	Source
1	Volume of patient	Patient	Objective	Systemic
2	Quality of in-house diagnostic centre—clinical laboratory			
3	Quality of in-house diagnostic centre—imaging			
4	Adequate beds			
5	Adequate quality—major surgical facility			
6	Adequate quality—minor surgical facility		Subjective	Systemic
7	Hygiene and cleanliness of the hospital	Physical Conditions		
8	Optimum utility facility—(water, electricity, drainage, sewerage)			
9	Adequate hospital supply of quality equipment			
10	Adequate hospital supply of consumables			
11	Availability of blood/ oxygen/ saline etc. life support			
12	Adequate number of non-clinical support staffs			
13	Adequate clinical support staff	Other Providers	Subjective	Systemic
14	Adequate physicians			
15	Governance—internal and external management	Governance	Subjective	Systemic
16	Self-efficiency	Self-Efficiency	Subjective	Individual
17	Efficiency of BMCH compared to other MCHs in the state of West Bengal	Comparison to MCHs	Subjective	Systemic

Source: Researcher's Consideration

5. Data source

We took Burdwan Medical Hospital as a case study and the eventually source for data. There are certain rationalities behind choosing BMCH as a case study among the eight MCHs in the state. The fundamental consideration is that it is a long-established MCH (more than forty-four years) and caters to a large hinterland of rural areas and urban centres. The urban centre of Burdwan has a large supply of private health care services, while the rural area starts within the surroundings approximately 5–7 km from the BMCH. Within the limits of the rural hinterlands there are also pockets of smaller urban centres which have some health care facilities both by public and private providers. Thus, BMCH is considered to be suitable for the study of the subjective parameters for efficiency analysis. We categorized service providers in four types: doctors, nurses, clinical staff and non-clinical staff. A total of one hundred samples were equally divided among the four categories of service providers. The heterogeneity in the sample is maintained through randomly chosen sample members in terms of "Years of Experience" and "Sex."

Table 9.2. Sample features

Providers	Sex		Total	Experience (years)	
	Male	Female		Minimum	Maximum
Doctor	16	9	25	1	31
Nurse	0	25	25	1	32
Clinical staff	21	4	25	5	21
Non-clinical staff	11	14	25	10	32
Total	48	52	100	1	32

Source: Researcher's Estimate

6. Empirical analysis

The findings of the survey results are presented through the following sub-sections, starting with the efficiency of the service state as perceived by the different service agents, followed by self-reported efficiency and agents' perceptions of efficiency of BMCH compared to other MCHs.

The service state as perceived by the different service agents is described by the marks obtained in the 10-point Likert scale. Minimum, maximum and average marks are presented to show the status of the service state. Standard deviation shows the variability in the efficiency judgment. We also estimated how skewed the efficiency judgment is.

We found that, to the doctor group, the service state is not efficient enough compared to the other three categories of providers. Among these four groups, the perception of efficiency is decreasing as the technical knowledge is increasing. A definite variability of opinions is also observed and opinions are skewed. The results are presented in Table 9.3 below.

The standard deviation of the self-reported efficiency among the doctors is highest compared to other service agents. All the four categories of agents judge the service state of BMCH as almost similar to the other MCHs in the state. However, there is intergroup variation in the judgment of service state efficiency among the agents.

In order to delve into the problem of understanding the range and effect of subjective efficiency, we resort to the regression analysis. We have selected three dimensions of subjective efficiencies: self-evaluation, relative evaluation of BMCH (in *pari passu* with other MCHs) and absolute efficiency as measured by the extent of overall performance. The dependent variables are chosen among the factors described in Table 9.3 above. In addition, we have considered the dummy variables to capture the fact whether the respondent is a doctor, a nurse, or a clinical or non-clinical worker. Since we are using subjective efficiency, these dummies are of paramount importance. However, the step regression process was used to identify the factors that are most contributory to the explanation of the subjective evaluations.

Table 9.3. Efficiency judgment of the service state BMCH *vis-à-vis* other medical college hospitals (MCHs)

Estimates	Service Agent-Doctors					
	Min.	Max.	Average	St. dev.	Highest Frequency	Skewness
Diagnostic Centre—Clinical Laboratory	1	5	2.6	1.2	2	0.53
Imaging	1	4	2.4	1	2	0.12
Adequate Clinical Support Staff	2	5	3.6	0.9	4	-0.05
Adequate Beds	3	6	4.1	0.9	4	0.56
Adequate Quality Major Surgical Facility	1	6	3.8	1	4	-0.38
Adequate Quality Minor Surgical Facility	1	4	2.9	1	3	-0.65
Adequate Physicians	4	8	5.6	1.3	5	0.34
Hygiene and Clinginess of the Hospital	1	5	3.5	1.3	5	-0.49
Optimum Utility facility	5	8	5.8	0.9	6	1.09
Adequate equipment	1	4	2.3	0.8	3	-0.14
Adequate Consumables	1	2	1.3	0.5	1	1.04
Adequate Non-Clinical Support Staff	3	8	5.6	1.4	6	0.03
Availability of Blood/ Oxygen/ Saline etc.	1	4	2	0.8	2	0.61
Governance—Internal and External Management	1	8	4.5	1.8	5	-0.15
Overall	1.9	5.5	3.6	1.1	3.7	0.2

Estimates	Service Agent-Nurses					
	Min.	Max.	Average	St. dev.	Highest Frequency	Skewness
Diagnostic Centre—Clinical Laboratory	5	7	5.8	0.7	6	0.3
Imaging	5	8	5.9	0.8	6	0.8
Adequate Clinical Support Staff	4	6	5	0.6	5	0
Adequate Beds	3	4	3.1	0.3	3	3.3
Adequate Quality Major Surgical Facility	4	7	5	0.8	5	0.4
Adequate Quality Minor Surgical Facility	6	9	7.6	0.7	8	0
Adequate Physicians	3	6	4.8	0.6	5	-1.4
Hygiene and Clinginess of the Hospital	2	5	3.4	0.7	3	0.2
Optimum Utility facility	6	8	6.8	0.7	7	0.3
Adequate equipment	4	6	4.9	0.6	5	0
Adequate Consumables	2	5	3.3	0.7	3	0.8
Adequate Non-Clinical Support Staff	4	7	5.1	0.6	5	1
Availability of Blood/ Oxygen/ Saline etc.	4	6	4.9	0.5	5	-0.2
Governance—Internal and External Management	1	9	5.6	2.8	7	-0.6
Overall	3.8	6.6	5.1	0.8	5.2	0.4

Estimates	Service Agent- Clinical Staff					
	Min.	Max.	Average	St. dev.	Highest Frequency	Skewness
Diagnostic Centre—Clinical Laboratory	5	9	7	1	8	-0.1
Imaging	6	9	4	0.8	8	0
Adequate Clinical Support Staff	4	6	5.1	0.5	5	0.2
Adequate Beds	3	5	4.2	0.5	4	0.3
Adequate Quality Major Surgical Facility	4	6	5.3	0.6	5	0
Adequate Quality Minor Surgical Facility	5	9	2	1	8	-0.4
Adequate Physicians	4	8	5.2	0.9	5	1.2
Hygiene and Clinginess of the Hospital	2	5	3.8	0.7	4	-0.3
Optimum Utility facility	4	7	5.7	1.1	7	-0.2
Adequate equipment	3	8	4.4	1.2	5	1
Adequate Consumables	3	5	3.6	0.6	3	0.6
Adequate Non-Clinical Support Staff	4	7	5.2	0.6	5	1.2
Availability of Blood/ Oxygen/ Saline etc.	5	8	5.9	0.9	5	0.6
Governance—Internal and External Management	1	10	5.6	2.8	9	-0.1
Overall	3.8	3	5.4	0.9	5.8	0.3

Estimates	Service Agent- Non-Clinical Staff					
	Min.	**Max.**	**Average**	**St. dev.**	**Highest Frequency**	**Skewness**
Diagnostic Centre—Clinical Laboratory	5	10	5	1.5	8	-0.3
Imaging	7	9	8	0.8	8	-0.1
Adequate Clinical Support Staff	5	9	2	1	8	-0.4
Adequate Beds	4	8	5.4	1	5	1.9
Adequate Quality Major Surgical Facility	7	10	7.7	0.7	8	1.3
Adequate Quality Minor Surgical Facility	8	10	8.6	0.7	8	0.7
Adequate Physicians	5	8	6	1	5	0.5
Hygiene and Clinginess of the Hospital	5	9	4	1.2	8	-0.5
Optimum Utility facility	8	10	8.4	0.6	8	1.3
Adequate equipment	7	9	4	0.7	7	1.2
Adequate Consumables	4	10	5.8	1.4	5	1.9
Adequate Non-Clinical Support Staffs	4	9	6.7	1.4	7	-0.2
Availability of Blood/ Oxygen/ Saline etc.	4	8	5.6	1.1	5	1
Governance - Internal and External Management	1	9	5.3	2.6	9	0
Overall	5.3	9.1	6.9	1.1	1	0.6

Estimates	Service Agent- Total					
	Min.	**Max.**	**Average**	**St. dev.**	**Highest Frequency**	**Skewness**
Diagnostic Centre—Clinical Laboratory	10	10	10	0	10	
Imaging	1	10	5.74	2.2	6	-0.5
Adequate Clinical Support Staff	1	9	5.93	2.4	8	-0.7
Adequate Beds	2	9	5.24	1.5	5	0.4
Adequate Quality Major Surgical Facility	3	8	4.2	1.1	4	1
Adequate Quality Minor Surgical Facility	1	10	5.46	1.6	5	0.2
Adequate Physicians	1	10	6.57	2.3	8	-0.9
Hygiene and Clinginess of the Hospital	3	8	5.43	1.1	5	0.7
Optimum Utility facility	1	9	4.52	1.9	4	0.8
Adequate equipment	4	10	6.68	1.3	6	0
Adequate Consumables	1	9	4.77	2	5	0.1
Adequate Non-Clinical Support Staffs	1	10	3.49	1.8	3	0.7
Availability of Blood/ Oxygen/ Saline etc.	3	9	5.64	1.2	5	0.8
Governance—Internal and External Management	1	8	4.59	1.8	5	-0.5
Overall	1	10	5.26	2.5	5	-0.1

Source: Researcher's Estimate

Table 9.4. Self-reported efficiency, overall and comparative efficiency of the service state

Category of Service Agent	Estimates	Self-Reported Efficiency	Status of BMCH Compared to Other MCHs	Overall Assessment of BMCH
Doctors	Min.	1	3	3
	Max.	8	5	4
	Average	6	4.1	3.6
	St. dev.	1.7	0.4	0.5
	Highest Frequency	7	4	4
	Skewness	-1.96	0.69	-0.43
Nurses	Min.	7	5	5
	Max.	8	6	6
	Average	1	5	5.1
	St. dev.	0.3	0.2	0.3
	Highest Frequency	7	5	5
	Skewness	3.3	5	3.3
Clinical Staff	Min.	7	4	5
	Max.	8	9	8
	Average	4	6.5	5.9
	St. dev.	0.5	1.2	0.8
	Highest Frequency	7	6	6
	Skewness	0.3	0.1	0.7
Non-Clinical Staff	Min.	8	3	5
	Max.	10	10	9
	Average	8.9	1	2
	St. dev.	0.4	1.6	1.3
	Highest Frequency	9	7	7
	Skewness	-0.8	-0.8	0
Total	Min.	1	3	3
	Max.	10	10	9
	Average	35	5.69	5.44
	St. dev.	1.4	1.5	1.5
	Highest Frequency	7	5	5
	Skewness	-1.7	0.6	0.5

Source: Researcher's Estimate

The first sets of regression are reported in Table 9.5a and Table 9.5b. The regression surprisingly shows that the categorical dummies have no role to play so far as the self-reported efficiency is concerned. Like the objective efficiency parameters under the subjective efficiency scheme the supply chain of infrastructure and supportive surgical and non-surgical items also become important. No service provider in the health sector can act without the adequate supply of necessary items. This partial correlation helps us to identify which of the significant factors is dominant. Under the vast set of adequacy of beds is the most dominant and significant explanatory factor. The result does not vary much whether we force or not the dummies within the step procedure.

The next important element is relative efficiency. According to our study the categorical dummies representing doctors and nurses become most important. It is to be noted that both these service classes are transferred from one hospital to another. As such, they have a wider experience of working in other MCHs. Thus, their evaluation becomes the most significant. The clinical as well as non-clinical staff have very little experience of other MCHs as service workers because they are generally ill-experienced. However, the number of non-clinical staff becomes another significant factor. The partial correlation analysis reveals the doctors' dummy as the most significant. As with the earlier case, the forcing or non-forcing of dummy variables make almost no difference to the nature of the relationships.

Lastly, we considered the absolute efficiency of BMCH or its overall evaluation. The variables that are significant here are some of those relative efficiencies. The only additional factor is the clinical staff dummy. Thus, there is very little difference in the pattern of relative and the absolute evaluation of BMCH. Again,thepartial correlation the doctors' dummy is the most dominant factor among those that are significant. The forcing and non-forcing of dummies again have no effect.

Table 9.5a. Factors responsible for difference in opinions—self reported efficiency (dummies forced in)

Stepped In Variables	Estimated Coefficients	Standard Error	T-ratio 96 DF	P-Value	Partial Correlation
Adequate beds	0.41558	0.8318E-1	4.996	0.000	0.454
Adequate quality minor surgical facility	0.24359	0.5964E-1	4.085	0.000	0.385
Adequate hospital supply of quality equipment	0.20431	0.7213E-1	2.832	0.006	0.278
Constant	3.0297	0.3884	7.800	0.000	0.623

R-square=0.6588; R-Square Adjusted=0.6481; n=100
Source: Researcher's Estimate

Table 9.5b. Factors responsible for difference in opinions—self reported efficiency (dummies forced out)

Stepped In Variables	Estimated Coefficients	Standard Error	T-Ratio 96 DF	P-Value	Partial Correlation
Adequate beds	0.41558	0.8318E-1	4.996	0.000	0.454
Adequate quality minor surgical facility	0.24359	0.5964E-1	4.085	0.000	0.385
Adequate hospital supply of quality equipment	0.20431	0.7213E-1	2.832	0.006	0.278
Constant	3.0297	0.3884	7.800	0.000	0.623

R-square=0.6588; R-Square Adjusted=0.6481; n=100
Source: Researcher's Estimate

Table 9.6a. Factors responsible for difference in opinions—MCH status of BMCH compared to others (dummies forced in)

Stepped In Variables	Estimated Coefficients	Standard Error	T-Ratio 96 DF	P-Value	Partial Correlation
Numbers of Non-Clinical Staffs	0.33726	0.8219E-1	4.103	0.000	0.386
Doctors Dummy	-2.5451	0.2370	-10.74	0.000	0.739
Nurses Dummy	-1.4632	0.2456	-5.958	0.000	0.520
Constant	4.7899	0.5082	9.425	0.000	0.693

R-square=0.6287; R-Square Adjusted = 0.6171; n=100
Source: Researcher's Estimate

Table 9.6b. Factors responsible for difference in opinions—MCH status of BMCH compare to other (dummies forced out)

Stepped In Variables	Estimated Coefficients	Standard Error	T-Ratio 96 DF	P-Value	Partial Correlation
Numbers of Non-Clinical Staffs	0.33726	0.8219E-1	4.103	0.000	0.386
Doctors Dummy	-2.5451	0.2370	-10.74	0.000	0.739
Nurses Dummy	-1.4632	0.2456	-5.958	0.000	0.520
Constant	4.7899	0.5082	9.425	0.000	0.693

R-square=0.6287; R-Square Adjusted=0.6171; n=100
Source: Researcher's estimate

Table 9.7a. Factors responsible for difference in opinions—efficiency of BMCH (dummies forced in)

Stepped In Variables	Estimated Coefficients	Standard Error	T-Ratio 96 DF	P-Value	Partial Correlation
Numbers of non-clinical staff	0.17964	0.7510E-01	2.392	0.019	0.238
Doctors dummy	-3.3988	0.2381	-14.27	0.000	0.826
Nurses dummy	-1.8326	0.2531	-240	0.000	0.596
Clinical staff dummy	-1.0613	0.2476	-4.286	0.000	0.403
Constant	6.0000	0.5258	11.41	0.000	0.760

R-square=0.7467; R-Square Adjusted=0.7360; n=100
Source: Researcher's estimate

Table 9.7b. Factors responsible for difference in opinions—efficiency of BMCH (dummies forced out)

Stepped In Variables	Estimated Coefficients	Standard Error	T-Ratio 96 DF	P-Value	Partial Correlation
Numbers of non-clinical staff	0.17964	0.7510E-01	2.392	0.019	0.238
Doctors dummy	-3.3988	0.2381	-14.27	0.000	0.826
Nurses dummy	-1.8326	0.2531	-240	0.000	0.596
Clinical staff dummy	-1.0613	0.2476	-4.286	0.000	0.403
Constant	6.0000	0.5258	11.41	0.000	0.760

R-square=0.7467; R-Square Adjusted=0.7360; n=100
Source: Researcher's estimate

7. Conclusion

Unlike the common belief, self-perceived efficiency is not heavily biased, showing considerable variation (except for non-clinical staff). Again, the availability of support materials is crucial for a positive self-evaluation. Infrastructural development is thus a pre-requisite for improving health-providers' efficiency.

References

Arrow, K. "Uncertainty and the Welfare Economics of Medical Care." *American Economic Review* 53 (1963): 941–969.

Basu Kaushik. "Beyond the Invisible Hand: Groundwork for a New Economics." Princeton: Princeton University Press, 2010

Rizzi Dino. "A utility-frontier production function approach for the efficiency measurement of multi-output public and non-profit organization: With an application to the academic departments of the University of Venice, Working." 2010. http://www.unive.it/media/allegato/DIP/Economia/Working_papers/Working_papers_2000/0008.pdf.

CHAPTER TEN

WES PRODUCTION INDEX FOR THE SECTOR SPECIFIC TOURISM STRATEGY: AN EMPIRICAL ANALYSIS OF SIKKIM

DEBASISH BATABYAL

1.Introduction

Sikkim is a small hilly state, bounded by vast stretches of Tibetan plateau in the north, the Chumbi Valley and the kingdom of Bhutan in the east, the kingdom of Nepal in the west and Darjeeling (West Bengal) in the south. Its latitude is $27^{0}30^{1}$ north and longitude $88^{0}30^{1}$ east. Sikkim is famous for scenic valleys, forests, snow-clad mountains, a magnificent Buddhist culture and heritage and its peace-loving people. Though small in area, the environmental, social and cultural diversities are not so. Sikkim has been given many names. The Lepchas, the original inhabitants of the land, called it Nye-mae-el, meaning "paradise." The Limbus named it Su Khim or "new house," while to the Bhutias it was Beymul Demazong, "the hidden valley of rice." Some scholars believe that the word Sikkim involves Nepalese dialect and refers to "a new place," or that the term is derived from a Sanskrit word meaning "mountain crest."

The people of Sikkim are ethnically diverse. The Bhutiascame from Tibet, the Lepchas were the aboriginal community and the Nepalese came from Nepal. When Sikkim was an independent state it faced many invasions by its neighbouring countries and the king took the help of British India, later gifting some of its region, including Darjeeling, to the British Indian Government. At present, this twenty-second Indian state (that merged with Indian Union in 1975) has over 81% of the total geographical area under the administrative managerial control of the Ministry of the Environment and Forests. Over 45% of the total geographical area of the state is under tree cover and nearly 34% of the geographical area is set aside as protected area in the form of national parks and wildlife sanctuaries. After merging with the Union, the rapid development ushered in a new era of tourism in Sikkim. The increased accessibility by roadways and air transport, rapid socio-economic development and competitive advantage both from the point of view of destination and geographical proximity to tourist generating states contributed to the development of tourism. Recognizing the increased tourist arrivals, accommodation units were set up in Gangtok and in a few towns mostly by outsiders without proper land use planning and architectural design.

The ever-increasing arrival of tourists and the dependence on tourism as a powerful industry and employment generating source, restoration of peace and harmony, hospitable people, plenty of diversified natural and cultural resources, and the typical interest of the people of India in discovering the unknown Sikkim brought about a new dimension for the development and marketing of tourism. The development and adoption of marketing extensively included a demand-supply equilibrium, and as such takes the tourism system into account. Destination marketing is unlike the marketing of essential consumables as it considers the characteristics of services. Again, in many cases the adoption and control is not fully devolved in corporate sectors. As a service marketing sphere it includes customer relationship management, internal marketing (with respect to a destination where all stakeholders are part of the product and contribute to the image and identity), the increased importance of strategic alliances/ linkages, etc. Destination marketing should be a part of destination management, but in Indian the concept of destinations is not in vogue and very often management and marketing are wrongly conglomerated with each other. There are very few destinations that are well managed by scientific research and background analysis. Sikkim has a poor quality database and only in a few cases are the available scientifically analyzed and interpreted, as the majority of destination planners and government officials are not from a tourism background. Therefore, an understanding of Sikkim as a popular destination along with the capacity levels (physical, biological, social, psychological and financial) largely contribute to the overall marketing strategy. The government of Sikkim is one of the very few Indian states trying to optimize benefits from tourism for their local people. The recently adopted policy to project the state as an Ultimate Ecotourism Destination was really a committed responsibility towards sustainable development; next to this are rural tourism and adventure tourism. Almost all these tourism facilities will contribute to the alternative tourism development in the state and the changes in types and forms will automatically have an impact on the activities of the tourists, duration of stay, the number of tourist arrivals and the tourism industry as a whole. So a proactive, environmentally-friendly approach has already been adopted in the marketing mix and Unique Selling Point (USP) of Sikkim, though there are many things still remaining. The major task for the government is how to coordinate and control interest of all participants and infuse the idea of sustainable practices in the state. The priority area is confused as there is a traditional clash between development and conservation. The destination marketing should not only increase the arrival of tourists but also be proactive in selecting target groups and introduce sustainable practices including mass awareness for environmental conservation.

The prime objective of the study is to introduce the concept of WES tourism production index in an Indian perspective as was used in the Alps region of Europe, and thereby measure it. Another objective is to measure the performance of tourism with a sector specific outline, and thereby formulate strategies. As regards the WES production index, the weight for each component in each category is considered with respect to the Indian and Himalayan State Sikkim. These weights are imposed after an opinion survey amongst industry leaders and consumers.

2. Methodology

Economists, business people and governments are often interested in finding out the rate of growth of certain economic variables, such as population, GNP, money supply, employment, productivity and trade deficit. Suppose we want to find the growth rate of personal consumption expenditure on services. Let Y_t denote real expenditure on services at time t and Y_0 denote the initial value of the expenditure on services. We may also recall the following well-known compound interest formula.

Here, the basic model is,

$$Y_t = Y_0(1+r)^t,$$

which on taking logarithm becomes

$$\ln Y_t = \ln Y_0 + t\ln(1+r)$$

Taking, $\beta_1 = \ln Y_0$ and $\beta_2 = \ln(1+r)$, the model can be rewritten as

$$\ln Y_t = \beta_1 + \beta_2 t$$

Adding the disturbance term u_t to the above model we have

$$\ln Y_t = \beta_1 + \beta_2 t + u_t$$

where, $U_t \sim N(0, \sigma^2)$

The WES Tourism Production index (T.I.) covers two aspects of the tourism industry: data collection for each and every sub-sector in tourism and measurement of performance (Bruges and hinterland area, West-Flanders region in Belgium 1962). The number and variety of the components was such to guarantee coverage of all measurable aspects of the tourism activity of the region. The index is given below:

$$\text{T.I.} = \sum_{j=1}^{n} \left| \sum_{i=1}^{n} [w_i \Delta comp.i] \right| w_j$$

$\Delta comp.i$ = change in component I with respect to reference period

w_i = weight component i within the corresponding category

w_j = weight of each category j.

The weights for three categories were 0.2, 0.5 and 0.3 for arrivals, accommodation and attraction, respectively. The accommodation sector was considered to be the most important as it indicates the core tourism infrastructure and actual demand for tourism. The weight for each component in each category was different. The weights for tourist arrivals (for different components) were 0.8 and 0.2 for domestic tourists and their international counterparts, respectively. The weights for accommodation units were 0.3 and 0.7 for standard accommodation and economy class accommodation, respectively. The room rent per person per night was Rs. 800 or more for standard accommodation, while rooms available at less than Rs. 800 were categorised as economy class. The third was attraction which had two associated weights of 0.75 and 0.25 for leisure and other components, respectively. All the category and component weights were based on the opinion survey among industry leaders and higher level government officials in Sikkim.

The study deals with the sub-sector wise analysis over a period of nine years (2002–2010) with an assertion that the performance of each sub-sector remains the same and policy issues are not required to change in the long term (five years, for example). So, the following hypothesis is given:

H0: There is no significant difference in the sub-sectors of tourism over the years as the important components of tourism remain the same with the arrival statistics.

3. Results and Discussion

Sikkim is politically divided into four districts, with the North (Mangan) being away from large market of West Bengal but with immense tourism potentialities. The capital city of Gangtok enjoys the development of infrastructure, modern civil amenities, access to health and education facilities and employment opportunities compared to all other three districts: South (Namchi) West (Gyashing) and East (Gangtok). Moreover, the North district is unfavourable in terms of its topography because of its extreme climate and mountainous terrain, which range in height from 17,000 ft to 28,000 ft. The towns are scattered, there are very few roads and the area is traversed by a single state highway. Only 7.69% of Sikkim's population live in this district, which account for 60% of land area. Its population density is consequently extremely low at only seven persons per sq. km. compared to 187,131 and 84 in the East, South and West districts, respectively. Even though its position has been improving, North Sikkim still scores lowest on both the Human Development and Gender Development Indices in recent years. In the North only 66.24% of the rural houses have electricity compared to 81.82% in the East, 77.78% in the West and 71.17% in the South (Gyatso & Bagdass 1998). As the arrival of tourism is one of the certain economic variables, its measurement of growth is imperative in the destination development literature. The most appropriate Log-Lin model was used to compute the growth rate of tourist arrivals.

For domestic tourist arrivals the model was derived to be

$$\ln Y_t = 9.59 + 0.11t$$

and the same for international tourists arrivals is

$$\ln Y_t = 7.37 + .084t\,.$$

Keeping in view the above equations with positive growth rates, it may be forecasted that the arrivals of both types of tourists will also increase for the upcoming years. The forecasted arrivals of tourists are shown in Table 10.1 below.

Table 10.1. Forecasted arrivals of tourists in Sikkim

Year	Arrival of Domestic Tourists (Forecasted)	Arrival of International Tourists (Forecasted)
2011[1]	493,856	20,537
2012	551,281	22,248
2013	615,383	24,101

Source: Authors own calculations based on the data published by the Sikkim Tourism Development Corporation, Govt. of Sikkim.

From the measure of Coefficient of Variation (CV) it may be noted that the domestic tourist arrival is fluctuating more, with its CV being 0.11 and the same for the foreign travellers being 0 .084.

Another study brought about a new orientation of tourism performance over the years with WES tourism production index. Though Sikkim is visited year round, the most popularly accepted Indian peak season[2] starting from October to March in every year has been taken into consideration to compare the tourism production index (T.I.) of Sikkim with that of the other Indian hilly states. As already indicated, this WES Tourism Production index covers two aspects of the tourism industry: data collection for each and every sub-sector in tourism and measurement of performance. The considered components cover all measurable aspects of the tourism activity of the region.

Two time periods, 2002–2003 and 2007–08, were considered the reference period and current period, respectively. Through an opinion survey of the different interest groups of the respective sub-sectors in tourism, weights for each category and component were set. These weights were fixed for the two periods. As Sikkim has controlled access and is dominated by roadways, arrival data through Rongpo was considered in terms of road transportation. Through opinion surveysin various sectors, e.g. accommodation, transportation and attraction, it is decided that hotels and/or supplementary accommodation units are the most important development indicators for tourism infrastructure, with the highest weight being 0.5. The second important aspect was attraction with its weight of 0.3 and roadways were placed in third position as they do not represent

[1] Arrival of foreign tourists and domestic tourists projected by the Horizon Study were 20,120 and 3,81,080, respectively, for 2011.

[2] During this season, Indian destinations experience the maximum number of foreign and domestic tourists every year.

tourism related infrastructure[3] directly and also include the transportation of hosts.

Considering the reference period and the current period, Table 10.2 below wasprepared and the values were put in relative form (i.e. absolute difference/actual value for the base year) with the attached weights for each category, i.e. j (e.g. arrivals, attractions, accommodation etc.) and component i (e.g. domestic, international, standard accommodation economy accommodation,[4] leisure etc.) respectively.

The study comprises a comparative assessment with respect to a base year or reference year 2002–03. Again, the study ensured a comparative assessment between two current years, i.e. 2003–04 and 2007–08, respectively. During the months of March and December in 2003–04 even negative growth rates were noted for *standard hotels* and the *purpose of travel for leisure*. These induced an alternative trend for the purpose of visit and/with non-standard accommodation, and induced an alternative strategy for tourism specifically for the winter season.

---

[3] Tourism Related Infrastructure is also called superstructure in tourism literature.

[4] More than Rs. 1,000 for standard accommodation, less than Rs. 1,000 for economy accommodation.

Table 10.2. Differences in annual growth rates

Months (Peak Season)[1] /Category	Arrival		Accommodation		Attraction	
	Domestic Tourists	International Tourists	Standard (Including Star Categories)	Economy	Leisure	Other Than Leisure
October	Y1=0.37 Y2=0.89167	Y1=0.75 Y2=1.1766	Y1=0.19 Y2=0.9148	Y1=0.33 Y2=0.9151	Y1=0.25 Y2=.6530	Y1=3.19 Y2=5.8921
November	Y1=0.39 Y2= 0.8072	Y1=0.74 Y2=1.2525	Y1=0.13 Y2=0.8371	Y1=0.50 Y2=0.8372	Y1=0.33 Y2=0.7353	Y1=1.12 Y2=1.7538
December	Y1=0.23 Y2= 0.7364	Y1=0.44 Y2=0.7326	Y1=0.36 Y2=0.7364	Y1=0.21 Y2=0.7362	Y1= -0.04 Y2=0.3890	Y1=2.71 Y2=3.8617
January	Y1=0.33 Y2= 0.120	Y1=0.60 Y2=1.1018	Y1=0.68 Y2=1.1897	Y1=0.26 Y2=1.1900	Y1=0.04 Y2=0.7762	Y1=3.02 Y2=4.9166
February	Y1=0.36 Y2=0.12917	Y1=1.03 Y2=1.0750	Y1=0.05 Y2=1.2789	Y1=0.49 Y2=1.2789	Y1=0.19 Y2=1.1448	Y1=1.61 Y2=2.0379
March	Y1=0.44 Y2=0.18811	Y1=0.61 Y2=0.6991	Y1= -0.12 Y2=1.7641	Y1=0.65 Y2=1.6742	Y1=0.30 Y2=1.6100	Y1=1.92 Y2=3.1467

Source: Authors own calculationsbased on the data published by the Sikkim Tourism Development Corporation, Govt. of Sikkim, 2010–11.

Notes: Reference Year 2002–03, Current Years (or Year 1 and Year 2) are 2003–04 and 2007–08 respectively.

[1] Following the conservative principle, this Indian peak period has been considered instead of a two-fold peak season comprising October to March and April to June.

Based on the values of the above-mentioned table, the WES indices are evaluated as in the following table.

Table 10.3. WES Tourism Production Index for Sikkim (peak period is October-March)

Month	Current Year (2003–04)	Current Year (2007–08)
October	0.53	1.23
November	0.44	0.91
December	0.38	0.89
January	0.50	1.37
February	0.44	1.30
March	0.52	2.06

Source: Author's own calculation based on the data published by the Sikkim Tourism Development Corporation (2010–2011)

Considering the values (in Table 10.4) of the final WES tourism production index, it is seen that these highlight overall massive growth during 2007–08, and the components of the index interrelate in such a manner that the index value is higher in every considered month than those in 2003–04. The index assumes a value greater than one in the months of October, January, February and March, and implies that the tourist arrivals, accommodation etc. start with an influx but dwindle with the advent of the winter season, and growing a little again with winter-specific tourists.

The earlier analysis is found to be contradictory in that the peak season[1] for domestic tourism in Sikkim is during the months of April to June, and after this the rainy season starts and the number of arrivals decreases. So, another alternative peak season was analyzed with the WES Index, given in Table 10.5 below.

[1] It is noteworthy that the peak season and lean season separately for domestic and international tourists were found in this study considering the basic arrival statistics from 2002 onwards.

Table 10.5. WES Tourism Production Index for Sikkim (peak period is April to June)

Month	Current Year (2003–04) w.r.t base year 2002–03	Current Year I (2005–06) w.r.t base year 2002–03	Current Reference Year(2007–08) w.r.t base year 2002–03	Current Year II (2010–11) w.r.t base year 2002–03
April	0.244	0.642	1.499	1.81
May	0.088	0.466	0.080	0.642
June	0.091	0.750	.0912	0.958

Source: Author's own calculation based on the data published by Sikkim Tourism Development Corporation (2010–2011)

Though the month of April is found to register an increasing number of domestic tourist arrivals over the years, a peak and trough is noticeable through the fluctuation in domestic tourist arrivals during the months of May and June. The fluctuating character of domestic tourist arrivals in Sikkim (which is the dominant form) may unhinge the supply side components of the destination and its management, andthis will alsohave an adverse impact on foreign tourist arrivals.

4. Conclusion

To analyze the sector-specific performance, the WES tourism production index has been used in the alpine Indian state Sikkim. This analysis has, surprisingly, shown a peak-trough in domestic tourist arrivals compared to international tourists. Sikkim is traditionally dependent on domestic tourists. However, a tendency of dwindling market demand with marked oscillations is already being noticed. The state is in dire need of increasing its foreign tourist arrivals to better shape its tourism sector with more income and international orientation. Following the tourism literature, the state is found to be in a vicious trap of cheap domestic tourism with unsustainable practices.

So, better tourism modules with all supply components need to be promoted with a new sustainable orientation of low volume and highly profitable tourism segments, such as alternative tourism, ecotourism, adventure tourism etc. The management of all resources and their allocation will have to revolve round this strategy. Accordingly, the types of tourism, types of tourists, spending patterns, future intention and retention studies, alternative routes and channeling of tourist traffic are found to be the core areas of development and management in Sikkim with its regional disparity in tourist arrival and varied development of supply components.

References

Allen M. D., Smith G. & Swanson J. R. “Tourism Policy and Planning: Yesterday, Today, and Tomorrow.” *UK, Elsevier*. 2008

Batabyal, D. “Significant Changes in Consumer Behaviour: A Case Study of Sikkim.” *Tourism Theory and Practice: Consumer Behavior Issue Kolkata* 8 (2) (2010): 50–57.

Batabyal, D. & Parida, B. B. “Review of Tourism Literature in a Destination Perspective: A Case Study of Sikkim.” *Tourism Theory and Practice: Tourism Literature Issue, Kolkata* 9 (1) (20101): 86–98.

Battacharya, B. “Tourism in the Himalaya in the context of Darjeeling and Sikkim.” In *Profiles in Indian Tourism*, Singh, S & Singh, T. (eds.). New Delhi, A.P.H. Publishing Corporation. 1996.

Buhalis, D. & Costa, C. (eds.). “Tourism Business Frontiers: Consumers, Products and Industry.” London: Elsevier, 2006.

Butler, R. “Modeling Tourism Development: Evolution, Growth and Decline.” In *Tourism Development and Growth*, Wahab S. & Pigram, J. J. (eds.). London & New York: Routledge, 2005.

—. “Problems and Issues of Integrating Tourism Development.” In *Contemporary Issues in Tourism Development*, eds. Pearce D. G. & Butler R., 65–79. New York: Routledge, Taylor and Francis e-Library, 2005.

Chettri N., Sharma E., Deb D. C. & Sundriyal R. C. “Impacts of Firewood Extraction on Tree Structure, Regeneration and Woody Biomass Productivity in a Trekking Corridor of the Sikkim Himalaya.” *Mountain Research and Development* 22 (2002): 150–158.

Diamond J. (2008). “Tourism's Role in Economic Development: The Case Reexamined.” *Economic Development and Cultural Change* 25 (3) (1977): 539–553.

Laws, E. “Tourist Destination Management: Issues, Analysis and Policies.” London: Routledge, 1995

Rahman, S. A.. “The Beautiful India-Sikkim.” New Delhi: Reference Press, 2006.

Sharpley, R. & Telfer D. J. “Tourism and Development in the Developing World.” London and New York: Routledge, 2008

Sharpley, R. "Tourism, Development and the Environment: Beyond Sustainability." UK: Earthscan, 2009

Southgate, C. & Sharpley, R. "Tourism, Development and the Environmentin Sharpley." In *Aspects of Tourism: Tourism and Development, Concepts and Issues*, eds. R. & D. J. Telfer, 231–262. Sydney: Channel View Publication, 2002.

Stamboulis, Y. "Destinations as Experience Stages: A Systems View." In *Tourism Development Revisited : Concepts, Issues and* Paradigms, eds. S. Babu. S, Mishra S., Parida B. B. London, Sage Publications, 2008.

Wahab, S. & Pigram, J. J. (eds). "Tourism Development and Growth: The Challenge of Sustainability." London & New York: Routledge, 2005.

Woodside A. & Martin D. (eds.). "Tourism Management: Analysis, Behavior and Strategy." UK: CEBI. 2007.

Vanhove, N. "The Economics of Tourism Destinations." London: Elsevier Butterworth Heinemann, 2005.

CHAPTER ELEVEN

HUMAN DEVELOPMENT INDEX AND GENDER INEQUALITY INDEX: RECENT CHANGES AND IMPLICATIONS

ANISH KUMAR MUKHOPADHYAY

1. Introduction

The Human Development Index (HDI) is a composite index aggregating three basic dimensions into a summary measure. This is published annually in the Human Development Report (HDR), using country-level information. The motivation behind the structure of the HDI was powerfully expressed in the 1990 HDR in the following terms.

Human development is a process of enlarging peoples' choices. In principle, these choices can be infinite and change over time. However, at all levels of development, the three essential ones are based on empowering people to lead long and healthy lives, to acquire knowledge and to have access to the resources needed for a decent standard of living. If these essential choices are not available, many other opportunities remain inaccessible. (UNDP 1990, 10).

The HDI has generated an extensive academic literature which has considered its properties, provided critiques and suggested potential improvements. With the occasion of the twentieth anniversary of the report, its authors undertook a comprehensive revision of these critiques and introduced several major changes in the 2010 edition. However, although this is not the first time the HDI has been modified, it is the first time some key changes have been introduced to the indicators used to measure progress and the functional form used to convert them to a single measure of progress.

Chakraborty (2011) has elaborately talked about the scope and choice of the indicators of HDI, arguing that the conceptual richness of human development cannot be totally captured by the three dimensions. He suggests that nutritional status, autonomy, mobility, freedom from crime and violence, and civil and political rights are such variables of utmost relevance, though they are absent from the index. Questions may be raised as to whether it is desirable to have a measure which is complete, scientifically perfect and logically correct or it is better to have a measure that may not be perfect but effective for advocacy and policy-making. It seems that the HDI is trying to maintain a balance between the two (Chakraborty 2011).

The purpose of this chapter is to discuss the new form of the HDI and the indices related to it. In the next section we shall talk about the methodology suggested by the UNDP to calculate the HDI, IHDI and GII. Then, in the following section, we will discuss their implications and limitations. The subsequent section is devoted to illustrating some preliminary findings. Finally, the chapter concludes with some comments.

2. Methodology

A number of significant changes have been incorporated into the UNDP-HDR-2010 in the formulation of the HDI. However, if we go back to the earlier reports and take stock since 1990, we find that over the periods there have been several changes in terms of HDI construction. The following table illustrates those changes.

Table 11.1. Pattern of HDI construction over the years

HDR	Bounds	Indicators: Health	Indicators: Education	Indicators: Income	Treatment of Income	Aggregation Formula
1990	Observed	Life expectancy at birth (UN Population Division)	Adult literacy rate (25+) (UNESCO)	Real GDP per capita PPP $ (log) (World Bank)	Logarithmic transformation with a cap	Arithmetic mean
1991 ~ 1993			(2/3)Adult literacy rate (UNESCO) (1/3)Mean years of schooling (UNESCO)	Real GDP per capita PPP $ (adjusted) (World Bank)	Atkinson formula with threshold value derived from poverty line	
1994	Fixed		(2/3)Adult literacy rate index (UNESCO) (1/3)Mean years of schooling index (UNESCO)		Atkinson formula with threshold value derived from global average	
1995 ~ 1998			(2/3)Adult literacy rate index (UNESCO) (1/3)Combined gross enrolment ratio index with a cap starting to bind in 1996 (UNESCO)			
1999				Real GDP per capita PPP $ (log) (World Bank)	Logarithmic transformation with a cap starting to bind in 2001	
2000 ~ 2009			(2/3)Adult literacy rate (15+) index (UNESCO) (1/3)Combined gross enrolment ratio index with a cap starting to bind in 1996 (UNESCO)			
2010	Upper: Observed Lower: Fixed		(1/2)Mean years of schooling index (Barro-Lee) (1/2)Expected years of schooling index (UNESCO)	Real GNI per capita PPP $ (ln) (World Bank)	Natural logarithmic transformation without a cap	Geometric mean

Source HDRO-RP-01

However, in HDR-2010 we have observed that the new formula is:

$HDI = (H_{Health} * H_{Education} * H_{Living\ standard})^{1/3}$
The Hi indices are still normalized indicators of achievements.

Life expectancy (le) remains the indicator for the health dimension, Gross National Income (gni) replaces GDP as the measure for living standards, while the mean years of schooling (mys) and expected years of schooling (eys) now make up the education dimension.

Hh = (le-lemin)/(lemax-lemin),
He = [((mys-mysmin)/(mysmax-mysmin)) *((eys-eysmin)/(eysmax-eysmin))]
Hls = (ln(gni)-ln(gnimin))/(ln(gnimax)-ln(gnimin)).

Broadly speaking, there are three steps for computing the IHDI. The first is to measure inequality in the underlying distributions. The IHDI draws on the Atkinson (1970) family of inequality measures and sets the aversion parameter ε equal to 1. In this case, the inequality measure is $A = 1 - g/\mu$, where g is the geometric mean and μ is the arithmetic mean of the distribution.

This can be written:

$$A_x = 1 - \frac{\sqrt[n]{X_1 \ldots X_n}}{\bar{X}}$$

where $\{X_1, \ldots, X_n\}$ denotes the underlying distribution in the dimensions of interest. A_X is obtained for each variable (life expectancy, years of schooling and disposable income or consumption per capita) using household survey data and the life tables. Note that the geometric mean in the above equation does not allow zero values, which requires specific adjustments in each dimension.

The second step is to adjust the mean achievement in a dimension, X–, adjusted for inequality as follows:

$$\bar{X}^* = \bar{X}(1 - A_x) = \sqrt[n]{X_1 \ldots X_n} \ .$$

Thus X–*, the geometric mean of the distribution, reduces the mean according to the inequality in distribution, emphasizing the lower end of the distribution.

The inequality-adjusted dimension indices, I_{IX}, are obtained as follows from the HDI dimension indices by multiplying them by $(1 - A_X)$, where A_X is the corresponding Atkinson measure:

$$I_{I_X} = (1 - A_X) \cdot I_X .$$

It should be noted that the inequality-adjusted income index I* is based on the unlogged gross national income (GNI) index, I*Income. This enables the IHDI to account for the full effect of income inequality, which in turn requires specific steps to re-estimate the HDI without logged income to enable the relevant comparisons to be made.

We can then compute the percentage loss to the HDI* due to inequalities in each dimension, which is calculated as:

$$Loss = 1 - \frac{IHDI^*}{HDI^*} = 1 - \sqrt[3]{(1 - A_{Life}) \cdot (1 - A_{Education}) \cdot (1 - A_{Income})} .$$

and the IHDI is then calculated as

$$IHDI = \left(\frac{IHDI^*}{HDI^*}\right) \cdot HDI$$

which is equivalent to

$$IHDI = \sqrt[3]{(1 - A_{Life}) \cdot (1 - A_{Education}) \cdot (1 - A_{Income})} \cdot HDI .$$

In HDR 2010, the Gender Inequality Index (GII) is a new measure of inequality faced by women and girls built on the same framework as the HDI and the IHDI to better expose differences in the distribution of achievements between women and men. The GII includes educational attainment, economic and political participation and female-specific health issues. Unfortunately, data limitations still significantly constrain the choice of indicators, although the indicators selected did allow application to 138 countries around the world. The GII is computed using the association-sensitive inequality measure suggested by Seth (2009). The index is based on the general means of different orders the first aggregation is by the geometric mean across dimensions; these means, calculated separately for women and men, are then aggregated using a harmonic mean across genders. One important feature of the GII method is its association-sensitivity; that is, unlike the IHDI, for example, we can take account of overlapping disparities. Aggregating across dimensions for each gender group by the geometric mean makes the GII association sensitive(Seth 2009). Following the technical notes of the UNDP-HDR, we now just report about the steps taken to construct the GII.

Step 1. Treating zeroes as extreme values, the maternal mortality ratio is truncated symmetrically at 10 (minimum) and at 1,000 (maximum). The maximum of 1,000 is based on the normative assumption that countries where the maternal mortality ratio exceeds 1,000 are not different in their abilities to create conditions and support for maternal health. Similarly, it is assumed that countries with 1–10 deaths per 100,000 births are essentially performing at the same level. The female parliamentary representation of countries reporting 0% is coded as 0.1%, because the geometric mean cannot have zero values and because these countries do have some kind of political influence by women.

Step 2. This involves aggregating across dimensions within each gender group using geometric means.

Aggregating across dimensions for each gender group by the geometric mean makes the GII association sensitive (Seth 2009).

For women and girls, the aggregation formula is

G_F=[(1/(MMR.AFR)$^{1/2}$(PR_F. SE_F) $^{1/2}$$LFPR_F$] ^{⅓}

G_M=[1 . (PR_M.SE_M) $^{1/2}$.$LFPR_M$] ^{⅓}

Step 3. This involves aggregating across gender groups, using a harmonic mean. The female and male indices are aggregated by the harmonic mean to create the equally distributed gender index

$HARM\,(G_F,G_M) = [(G_F)^{-1} + (G_M)^{-1}/2]^{-1}$

Step 4. This involves calculating the geometric mean of the arithmetic means for each indicator. The reference standard for computing inequality is obtained by aggregating female and male indices using equal weights (thus treating the genders equally) and then aggregating the indices across dimensions:

$G_{F,M}$=[*Health . Empowerment. LFPR]*^{⅓}

Where Health=[1/(MMR.AFR)$^{1/2}$ +1]/2

Empowerment=($\sqrt{}$*PRF . SEF* + $\sqrt{}$*PRM .SEM*)/2

LFPR=($LFPR_F$+$LFPR_M$)/2

Health should not be interpreted as an average of corresponding female and male indices, but as half the distance from the norms established for the reproductive health indicators—fewer maternal deaths and fewer adolescent pregnancies.

Step 5. Calculating the Gender Inequality Index comparing the equally distributed gender index to the reference standard yields the GII
$1 - Harm\ (G_F, G_M)/\ G_{\mathrm{F,M}}$

3. Implications and Limitations

The HDI has been severely criticised during its early days on the grounds of choice of variables, functional form, redundancy and robustness, dimensions and weights, indicators and country classification. However, after its reconstruction in 2010, some new frontiers of criticism have emerged. This is a research paper by the HDRO-RP/01 where the authors have discussed the existing shortcomings of the new HDI and related constructional problems, which will be explored below.

According to the HDRO-RP/01, Herrero, Martinez & Villar (2010) talk about the numerous problems with the old HDI and show the possible ways out. Among them, one is switching to the geometric mean, encompassing the issue of inequality into the dimension of living standards, and changing indicators for the health and education dimensions. In fact, the new HDI is similar to the formula previously suggested by Herrero, Martinez & Villar (2005). Zambrano (2011a) provides an axiomatic characterization of the new HDI, showing that a number of desirable axioms are satisfied, namely monotonicity, subsistence, independence and basal growth. Zambrano (2011b) concludes that: "… The new methodology chosen by the HDRO in 2010 is a vast improvement over both the old formulation and the numerous alternatives that at one point or another have made it to the drawing board." However, criticisms can still be observed, as set out in the following.

(a) They also mentioned the debate initiated by Martin Ravallion (2010) regarding the trade-offs inherent in the new HDI, which is a cause of apprehension. Ravallion obtained the marginal rates of substitution between longevity and income in the new HDI (MRSh,y) and shows that these vary from $0.53 per year of life in Zimbabwe to about $9,000 in the richest countries, asserting that "the highest valuation of longevity is 17,000 times higher than the lowest." Ravallion draws the policy implication that "if one accepted the trade-offs embodied in the new HDI, one would be drawn to conclude that the most promising way to promote human development in the world would be by investing in higher life expectancy in rich countries—surely an unacceptable implication of the HDI's trade-offs." He also recommends an alternate functional form which he attributes to Chakravarty (2003) and claims could have avoided this feature of steep increase of the "value" of longevity with per capita income.

(b) According to Klugman et al. (2011) there are several major problems with Ravallion's line of argument which need to be addressed because they have attracted some attention. Klugman et al.(2011) suggests that Ravallion's view is wrong in his assertion that the new formula has intensified this problem. His evidence for this assertion is a comparison of the $MRS_{h,y}$ in the new and old indices. However, he ignores the fact that incomes were restricted at \$40,000 in the old index. Formerly, $MRS_{h.y}$ was a constant fraction of income for incomes lower than \$40,000, and infinity for incomes higher than or equal to \$40,000. This discontinuity is key as it implies that the old version of the index was much more concave regarding income than the new version. It also implies that the ratio between the "value" of longevity in the richest to the poorest countries in the old HDI was infinity, contrasting the new formula, where it is a positive real number. However, there are more basic issues relating to why Ravallion's concerns are misplaced.

(c) Klugman et al. (2010) discuss why an index of capabilities is conceptually different from a social welfare function(SWF). The key difference is that a SWF is intended to be maximized, and thus the trade-offs along that SWF can be interpreted as values. However, a capability index is meant to give a measure of the extent to which people in different countries have access to substantively different lives. Obviously, we care about the expansion of these capabilities and believe that expansion is welcome. But this is entirely separate from arguing that the maximization of capabilities should be the only objective of social action. There are reasons why it should not recommend that societies try to maximize the HDI, and there are components of human development or well-being that are not included in the HDI. Maximizing the HDI would imply attributing a zero weight to the excluded dimensions, completely going against the spirit of the human development approach which has forcefully argued for the relevance of the broader dimensions in human development.

(d) Expansion of capabilities is only one of the reasons why people may care about the components of the HDI. People may enjoy the luxuries that come from higher levels of income even if these don't contribute to their leading substantively different lives. For this reason, policymakers can and often do put weight on the generation of economic growth in rich countries that goes beyond anything justifiable if they only care about the expansion of the capabilities derived from that income. From the standpoint of the human development approach, there is nothing inherently wrong with this, and recognizing the value of capabilities is not the same thing as saying that they should be the one and only overriding concern.

(e) Moreover, the HDI is a crude measure of development and is severely limited by data constraints, and thus one can at best expect it to give a rough characterization of a country's relative stance. Policymakers in any particular country are likely to have superior information to inform their decision-making process at their disposal. The MRS is better interpreted as reflecting the differing relative contributions of income *vis-à-vis* health in furthering capabilities. Since the relative contribution of income to furthering capabilities is infinitesimally small in rich countries, it appears to be a large number by comparison in poor countries. Hence, the 17000:1 ratio purely suggests that the HDI puts a very low weight on increases in income in rich countries compared to the very high value on health improvements. Because of this, improvements in longevity matter much more than improvements in income to the development of capabilities in rich countries and the implied valuation of longevity in a rich country is enormous. In other words, the 17,000:1 ratio only tells us that in rich countries, income is nearly valueless in terms of enhancing capabilities. This, in fact, is a simple implication of the idea that beyond a certain level income contributes very little to the expansion of capabilities. Putting it differently, and dealing with the MRS point more directly, if the HDI measures some kind of country-level welfare, the fact that Zimbabwe exhibits a $MRS_{h,y}$ (i.e. a marginal value of health in terms of dollars) of only 53 cents, while Qatar has one that is $8,800, simply reflects the relative abundance of health compared to income in Zimbabwe, and the relative abundance of income compared to health in Qatar.

If we say that longevity is worth 17,000 times as much in Qatar as in Zimbabwe, then that will possibly lead some readers to deduce that added (e.g. aid) dollars should go to Qatar instead of Zimbabwe. This presumes that the dollars Qatar possesses should be spent on health improvements, but the dollars that Zimbabwe has should be directed toward enhancing people's livelihoods. However, whether the ratio has any broader significance is hard to discern from the simple MRS computation.

Ravallion advocates dropping the logarithmic transformation of income so that the gradient of the valuation of longevity in income would be less steep. The problem gets bigger in this case. The HDI is a convex function in capabilities, and the dimension index Hi, which measures the capability of "a decent standard of living," is concave in income (as it uses the logarithmic transformation). The first of these features comes from the idea that there should be imperfect substitutability in capabilities, whereas the second comes from the idea that the capability of a decent standard of living is a concave function of income. The asymmetric treatment of income has strong roots in the capability theory. The authors of the early HDRs argued that if HDI is an index of capabilities, then income should be treated differently. This is because income is not a capability but rather an input that people use to develop and use these capabilities. It is a means, not an end in itself. When income is low, its growth helps people to gain access to essential goods and services which constitute relevant capabilities. Yet as income rises its contribution to further expanding capabilities declines. Therefore, one must draw a distinction between the input into the formation of capabilities (income) and the resulting capability (a decent standard of living). The fact that income turns into capabilities at a declining rate implies that the function mapping income into capabilities should be concave. Dropping the log transformation, however, would have been equivalent to disregarding this key distinction between capabilities and inputs into the formation of capabilities. The idea that income is not a capability and thus requires an asymmetric treatment compared to that given to health and education, appears to have been well understood by Chakravarty (2003), who developed a generalized family of concave indices that generalized the HDI, all of which used the concave piecewise function of income designed in the early Human Development Reports to capture the idea that income turns into capabilities at a declining rate (a function which was later replaced by the logarithmic transformation). Ironically, Ravallion claims to apply the Chakravarty formula without using any concave transformation of income.As shown by Zambrano (2011b), the real Chakravarty index (i.e. the index using the concave transformation of income as originally posited by Chakravarty) actually delivers rankings which are much closer to those given by the new HDI than to those of the Ravallion index.

Now let us have a look at the GII construction and the problems associated with that. Permanyer (2011) provides a critical assessment of the new GII presented in the HDR-2010. Clearly, the new index is an important contribution to the debate on gender inequality measurement. On the one hand it incorporates important reproductive health variables that were not previously used in a global gender inequality index, and on the other hand it proposes a new methodology to aggregate multidimensional information into a single dimensional index. However, he has identified two major flaws in the construction of the index for which some remedy might be sought. To start with, the functional form of the index is not particularly user-friendly. While the choice of that functional form is coherent with the methodology used in the construction of the new HDR measures (i.e. the new HDI and IHDI), the GII has been unnecessarily complicated to satisfy certain normative properties that are otherwise satisfied by much simpler indices. Moreover, the GII incorporates both women-specific indicators (i.e. absolute) and woman vs men indicators (i.e. relative) into a single formula. This creates a host of important problems, both conceptually and methodologically. Conceptually, the mixture of absolute and relative indicators makes the indices difficult to understand and the interpretation of an already complicated index becomes even more incomprehensible. This choice produces an index that, among other things: (i) unjustly penalizes less-developed countries for poor performances in reproductive health indicators that are by no means entirely explained by the gender-related norms or discriminative practices against women that the GII purports to measure, and (ii) does not reach the expected value(or even normatively desirable value) whenever women and men fare equally in all indicators. He has offered some alternatives to the identified shortcomings. For instance, Beneria & Permanyer (2010) recently proposed a much simpler Women Disadvantage Index (WDI) by substituting the complicated GII methodology. WDI is just an average of the gender gaps in which men outperform women. Moreover, this index only uses relative indicators that compare male and female achievement levels, thus avoiding the many problems assailing the GII. In order to substitute the women-specific components of the GII (MMR and AFR), the WDI incorporates gender-specific life expectancies at birth, an indicator that has been widely used in previous global gender inequality assessments. It turns out that WDI is weakly and positively correlated with the GDP per capita as opposed to what happens with the GII, which is biased in favour of richer countries thus illustrating its success in the task of capturing new information that is not put in a nutshell in the GDP. As opposed to GII, WDI has the further advantage of being decomposable by subcomponents, thus facilitating the understanding of the internal structure of the index and allowing easily computation of the percentage contribution of each individual subcomponent to the aggregate value of the index. Using data from the 2010 HDR they find that the values of WDI are strongly dominated by the gender gap in parliamentary representation—an indicator that has been criticized elsewhere for excluding political participation at the community and local levels.

4. Findings and Results

The empirical exercises have been done in two stages. To start with we have calculated Spearman's product moment correlation coefficient among the chosen indicators. It is evident from Table 11.2 below that the degree of co-variation has only increased among the respective variables in 2011 compared to 2010. In each case, the value has risen and the level of significance is quite high (5%). Also, the nature of co-variation has the negative sign, which is quite expected. For instance, net HDI will always be inversely proportional to the GII, which suggests that quality of life in the basic health and education sector gets diminished if gender bias in each parameter rises. The same is true for IHDI and GII association. However, GII falls with an improvement in GNI, and this is true because a reduction in the value of GII means a decline in the level of inequality.

Table 11.2. Pattern of correlation coefficients during 2010 and 2011

Variables	R_{10}	R_{11}
GII, NHDI	-0.8720*	-0.9003*
GII, GNI	-0.8633*	-0.8383*
GII, IHDI	-0.9240*	-0.9415*
NHDI, GNI	0.7744*	0.7860*
NHDI, IHDI	0.9666*	0.9693*
IHDI, GNI	0.8475*	0.8585*

* implies level of significance at 5%

In the second stage of the empirical exercise we want to see how measures based on health, education and gender bias in several sectors are associated with income. Here, NHDI has been derived from the HDI where we have simply deducted the income component from HDI and obtained the GM of the health and education index. NHDI therefore implies net HDI. GII/NIHDI is a ratio that essentially captures the bias in terms of gender where NIHDI is the net Inequality adjusted HDI value and derived inthe same way as NHDI. The basic regression equations that need to be estimatedare given below:

$NHDI=a_1+a_2 Ln(GNI)+u_1$...... (1)
$GII=b_1+b_2 Ln(GNI)+u_2$..........(2)
$GII/NIHDI=c_1+c_2 Ln(GNI)+u_3$.... (3)
$GII=d_1+d_2 Ln(NHDI)+d_3 Ln(GNI) +u_4$..... (4)

We have collected data for two years at a cross-country level where we have 110 countries in each year. However, if we pool them together for two years then the number of countries becomes 135. The estimation has been made with the help of the OLS method. The error components u_1 u_2 u_3 u_4 are normally distributed. In each equation, the association has been observed with the income level. For example, analysis has been made to focus on how NHDI is related in each year with LnGNI and if there is any change when the data are pooled. However, in equation (4) the effect of income and NHDI has been observed on GII.

In Table 11.3 below we have shown the regression results of all equations. If we take a look at the estimates of the first equation then in each case we observe that the association is highly significant with a very good value of Adj-R^2. In the second equation the observation is more or less the same with the difference that here the association is inverse and the Adj- R^2 is lower than the previous case. The implication is that with an increase in income the value of GII should come down, which means the gender bias will be lower.

In the third equation the ratio of GII/NHDI, which is also a measure of gender bias relative to net HDI, is negatively related. Evidently, the relationship is highly significant with the quite substantial value of Adj-R^2, but its value is lower in the case of pooled data.

Equation four is actually a joint association of Ln(NHDI) and Ln(GNI) when estimated against GII. Now as Ln(NHDI) increases, GII falls and they are therefore negatively related. The pattern of association is similar in the case of Ln(GNI) and GII. Like before, the t-statistics are highly significant with a quite robust Adj- R^2.

However, excepting the first case, in each of the equations we find that the value of Adj - R^2 in the case of the pooled estimate is less than its corresponding cross-section values, although other things have remained the same. This is interesting because we normally expect it to be better in the case of pooled data.

Table 11.3. Regression analysis during 2010–2011

Variable	Indep Variable	Coefficient			t-Statistic			P-Value			Adj-R^2		
		2010	2011	Pooled	2010	2011	Pooled	2010	2011	Pooled	2010	2011	Pooled
NHDI	CONS	-.328	-.336	-0.298	-6.92	-7.46	-8.63	0.000	0.000	0.000	0.761	0.772	0.762
	LnGNI	.117	.121	0.113	21.97	23.80	29.41	0.000	0.000	0.000			
Dep GII	CONS	1.603	1.51	1.455	26.67	25.41	26.23	0.000	0.000	0.000	0.649	0.66	0.546
	LnGNI	-.120	-.129	-0.111	17.90	-19.02	-18.02	0.000	0.000	0.000			
GII/ NHDI	CONS	4.098	3.489	3.710	24.74	23.37	29.06	0.000	0.000	0.000	0.776	0.772	0.666
	LnGNI	-.364	-.325	-0.330	-19.57	-19.33	-23.20	0.000	0.000	0.000			
GII	CONS	1.21	.947	.652	8.53	6.57	5.76	0.004	0.000	0.000	0.760	0.798	0.630
	LnGNI	-.085	-.076	-.038	-6.29	-5.46	-3.56	0.000	0.000	0.000			
	LNNHDI	-.191	-.290	-.408	-2.93	-4.29	-7.90	0.000	0.000	0.000			

5. Concluding comments

We have sketched out the formation of the HDI, IHDI and GII the way UNDP has made them. We have then tried to analyse why and how these measures are better placed than the previous ones. Very briefly, we have surveyed the criticisms of the old HDI and explained how the new HDI has addressed them. We have shown that some of the criticisms of the new functional form stem from a basic misunderstanding of the conceptual basis of the HDI. There is still substantial room for progress in the measurement of human development. The 2010 HDR made a significant move away from the idea that the ideal measure of human development must cover only the three core dimensions, and presented three new measures that take different aspects of the distribution of human development into account.

While the GII is an interesting and novel way of conceptualizing gender inequality, it has been argued that the particular way in which the index was constructed limits its usefulness and appropriateness as a global gender inequality index. In particular, it has been contended that the functional form of the index is excessively and unnecessarily confusing. Moreover, the inclusion of indicators that compare the relative performance of women *vis-à-vis* men together with absolute women-specific indicators further obscures the interpretation of an already complicated index.

From the empirical standpoint an attempt has been made to investigate how income is playing a role in relation to the non-income component of human development and gender bias. As expected, the associations are quite healthy in terms of the level of significance. However, it will be interesting to see the same kind of associations with the old HDI. It has been observed that IHDI is a very useful measure by which both inequality and development can be observed across countries. The role of income is very crucial in this analysis, and how an income expansion helps develop the performances in health, education, empowerment of women etc. will be interesting to investigate using more data points in a cross-country framework.

References

Anand. S & A. K. Sen "Human Development Index: Methodology and Measurement." *Human Development Report Office Occasional papers No.12*, New York: UNDP, 1994.

Beneria, L. & Permanyer, I. "The Measurement of Socio-economic Gender Inequality Revisited." *Development and Change* 41 (2010): 375–399.

Chakravarty, S. R. "A Generalized Human Development Index" *Review of Development Economics* 7 (1) (2003): 99–114.

Chakraborty, A. "Normative Measurement." An unpublished paper given at the refresher course in Jadavpur University. 2011

Chatterjee, S. K. "Measurement of Human Development—An Alternative Approach." *Journal of Human Development* 6 (1) (2005): 31–53.

Gaye, A., Klugman, J., Kovacevic, M., Twigg, S. & Zambrano, E. "Measuring Key Disparities in Human Development: The Gender Inequality Index." Human Development Research Paper 46. UNDP–HDRO, New York. 2010.

Herrero, C., Martínez, R. & Villar, A. "A Multiplicative Human Development Index." Working Paper, Fundación BBVA / BBVA Foundation. 2005.

—. "Improving the Measurement of Human Development." Human Development Research Paper 12, UNDP–HDRO, New York. 2010

Klugman, J., Francisco, R. & Choi, H. "The HDI 2010: New Controversies." Old Critiques Human Development Research Paper 2011/01. 2011

Permanyer, I. "A Critical Assessment of UNDP's Gender Inequality Index." Unpublished research paper. Center for Demographic Studies, Barcelona, Spain. 2011

Ravallion, M. "Mashup Indices of Development." Policy Research Working Paper 5432, World Bank, Washington DC. 2010a

—. "Troubling Tradeoffs in the Human Development Index." Policy Research Working Paper 5484, World Bank, Washington DC. 2010b

Seth, S. "A Class of Association Sensitive Multidimensional Welfare Indices." OPHI Working Paper No. 27. 2009a.

Seth, S. "Inequality, Interactions and Human Development." *Journal of Human Development and Capabilities* 10 (2009b): 375–396.

Ul.Haq, M. "The Human Development Paradigm." In *Readings in Human development*, eds. Parr, S. F. & A. K. Shiva Kumar. Oxford: Oxford University Press, 1995.

UNDP (United Nations Development Programme). "Human Development Report 1990, 2010, 2011." New York: Oxford University Press.

CHAPTER TWELVE

ROLE OF TRADE POLICIES IN A DYNAMIC MODEL OF CHILD LABOUR IN A CHOICE-THEORETIC FRAMEWORK

SOUMYA SAHIN AND AMBAR NATH GHOSH

1. Introduction

The problem of child labour is a major challenge to the progress of developing countries. Children work at the cost of their right to education, as a result of which they are forced to work as unskilled labourers even when they are adults. The distributional problem is accentuated by the fact that there is growing intra-country wage inequality between skilled and unskilled workers in almost every part of the globe. The situation is particularly grave in Africa and South-East Asian countries. In its global report on child labour, the International Labour Organization (ILO) recorded that in 2008 there were almost 215 million children around the world who did regular work.

There has been an extensive literature on how the problem of child labour in developing countries can be tackled effectively. Though the existence of child labour is a social malaise, coming up with a straightforward hypotheses may not be the right approach. To address this issue and suggest policy prescriptions, we first need to understand why child labour exists at all. Economists seeking government intervention in the child-labour market often put forward the logic that there are externalities to child labour since private returns to education are smaller than social returns. A ban on child labour leads to more resources allocated to children's education, brings the average level of human capital closer to the optimal, and consequently increases welfare. Udry (2006), Baland & Robinson (2000), Ranjan (2001), Hazan & Berdugo (2002), Fan (2004a; 2004b) and Doepke & Zilibotti (2005) examine the relationship between child labour and human capital accumulation.

In the popular mind, child labour is very often equated with child abuse. The phenomenon is taken to be a product of profit-seeking entrepreneurs looking for cheap labour and selfish parents who prefer to send their children to work to enjoy more consumption. There are bargaining models which analyze intra-household decision-making where parents and children are involved in bargaining conflicts (Moehling 1995; Gupta 1998). On the other hand, there are more conventional models which assume parental altruism and thereby take the position that when we have children working as a mass phenomenon, as in many less-developed countries, it is much more likely that this reflects not a difference in the attitude of the parents but the problem of stark poverty. Because of this, parents are compelled to send their children to work for reasons of survival (Kambhampati & Rajan 2004; Edmonds & Pavcnik 2005). Soares (2010) presents an overlapping generation model where parental altruism results in transfers that children allocate to consumption and education. A ban on child labour decreases children's income and generates an increase in parental transfers, bringing their levels closer to the optimum and raising children's welfare, as well as average welfare, in both the short and long terms. Increase in transfers increases the level of human capital, and leads to a fall in savings and hence physical capital accumulation. When prices are flexible these effects diminish the positive welfare impact of the ban on child labour. Dessy & Pallage (2005) present an environment where although child labour has a direct negative impact on children's well-being, it may be the best available choice. Hence, a ban on child labour deprives children of their best available choices and can make them worse off. Basu & Van (1998) explore the possibility of multiple equilibria in two versions of the labour market—one in which wages are low and children work, and the other in which wages are high and children do not work. As a result of a total ban on child labour, the "bad" equilibrium may cease to exist and the economy settles into the "good equilibrium." This is precisely a consequence of multiple equilibria. Dessy & Pallage (2001) also derive a similar conclusion.

Although there is a significant body of literature on the economics of child labour, relatively speaking, only a few studies have formally addressed its international economics. Dinopoulos & Zhao (2007) use a standard general equilibrium two-sector model of a small open economy to show that a ban on child labour benefits unskilled adult workers but hurts skilled workers. Maskus (1997) has developed a two-sector, specific-factors model, where child labour is modelled as a specific factor employed in the exportable sector and adult labour is modelled as a mobile factor. Trade liberalization raises the output of the exportable sector and increases the demand for child labour and child wages. Ranjan (2001) analyzes the effects of trade liberalization on the return to education in the presence of credit constraints. Credit-constrained families can withdraw their children from the labour force when the family's income reaches a minimum level. Trade openness affects a poor, unskilled labour-abundant country by raising the return to unskilled labour and inducing parents to educate their children instead of sending them to work.

Meanwhile, the nexus between trade openness and credit constraints is the focus of Jafarey & Lahiri (2002), who developed a 2-period, two good model with skilled and unskilled labour. Children can acquire skills through training instead of working. In their model, poor families (headed by unskilled parents) choose less education than rich families (headed by skilled parents). Easier access to international capital markets reduces the interest rate and increases the return to education. As a result, the incidence of child labour is reduced in this case. Trade liberalization raises the price of non-skill-intensive goods, reduces the returns to education, and could lead to an increase in the incidence of child labour. Basu & Chau (2004) analyze the effects of trade openness in a dynamic model of child labour and debt bondage. Trade openness increases the short-term supply of child labour but does not affect it in the long term.

Doepke & Zilibotti (2010) have analyzed the effect of international labour standards and its impact on the incidence of child labour in a political economic framework. They have come out with the conclusion that these types of international action might be counter-productive. Therefore, the policy-makers have to be careful about the fact that such well-meaning intervention may be used by certain lobbies with other agendas, such as protectionism.

However, the models discussed above are all based on the assumptions that parents are either self-interested and child labour occurs as a result of higher bargaining power of parents *vis-à-vis* children, or parents are altruistic but child labour still exists because the adult wage is so low that it cannot cover the subsistence needs of the family. However, in our model we present a theoretic framework where parents compare the utility derived from their child's income if they send them to work to the utility they derive from their education. The production side of the model consists of an export sector which employs only adult unskilled labour and child labour. Following Basu & Van (1998), Ranjan (2001), Doepke & Zilibotti (2005) and Genicot (2005) we adopt the standard assumption that adult unskilled labour and child labour are substitutable. Also, there is an import-competing sector which employs both skilled and unskilled labour. It is a small open economy with a tariff rate, τ, imposed on imports and all the markets are perfectly competitive. In section 2 of the model we determine the equilibrium wage rates and the size of the two sectors. In addition, we see the effects of a change in tariff rates on the size of the two sectors.

2. The model

We develop a simple overlapping generations model for a small open economy. There are only two groups of economic agents: households and firms. The production sector is divided into two segments: an export sector and an import competing sector. The former produces only for the export market, while the latter caters only to the domestic market and its output competes with imports.

2.1 The households

Individuals have two-period lives. Each parent has one child. There is a continuum of individuals in every generation over the interval $[0, N]$, so that the size of each generation is *N*. Parents send their children to school if and only if they consider it optimal to do so. Children who go to school work only in the second period of their lives. Those who do not go to school work in the first period in child labour and in the second period in unskilled adult labour. Child labour and unskilled adult labour are homogeneous except that the former supplies less labour than the latter in any given period. We assume that an adult worker (skilled or unskilled) inelastically supplies one unit of labour in any given period and a child supplies aβ fraction of the labour supplied by an unskilled adult in any given period. Each household in every period consists of a parent and a child. The parent is the decision-making unit, and altruistic in the sense that they derive utility from sending their child to school as well. They do not save any part of their household income, and decide whether to send their child to school or not. This is the only problem they face. To take this decision, they compare the utility they derive from their child's income if they send them to work to the utility they derive from their education. The wage rate of child labour in period t is W_t^C. The utility a parent derives by sending their child to work in period t is $U(W_t^C)$. A child acquires a given level of skill, denoted $\overline{H}$, from their education. The utility that a parent derives from their child's education is $U(\overline{H})$. We assume that $U(W_t^C)$ is identical for every parent, but $U(\overline{H})$ varies across parents. Every parent has a unique $U(\overline{H})$. $U(\overline{H})$is continuously distributed over the set $[0,N]$for the parents of every generation. We assign to each parent a number, i, equal to $U(\overline{H})$so that i is also continuously distributed over $[0,N]$. The situation is shown in Fig. 12.1 below, where $U(\overline{H})$is measured on the horizontal axis, while the numbers assigned to parents are measured on the vertical axis.

Regarding $U(W_t^C)$, we ignore the diminishing marginal utility of income, which is not important in the present context, and assume for simplicity that

$$U(W_t^C) = W_t^C \qquad (1)$$

A parent will send their child to school if and only if$U(W_t^C) \leq U(\overline{H})$, i.e. if and only if

$$W_t^C \leq U(\overline{H}) = i \, (2)$$

At the margin the following equality holds:

$$W_t^C = i \, (3)$$

From (3) it follows that in period t, all the parents with numbers less than W_t^C in the interval $[0,N]$send their children to work, while the remaining parents with numbers higher than W_t^Cin the interval $[0,N]$send their children to school. The child labour number in period t is therefore W_t^C, provided $W_t^C \leq N$, which we assume to be the case here. The situation is shown in Fig. 12.2, where the equilibrium number of child labour corresponds to the point of intersection of the horizontal W_t^C line and the 45^0 line.

Fig. 12.1. Parents' utilities from children's education

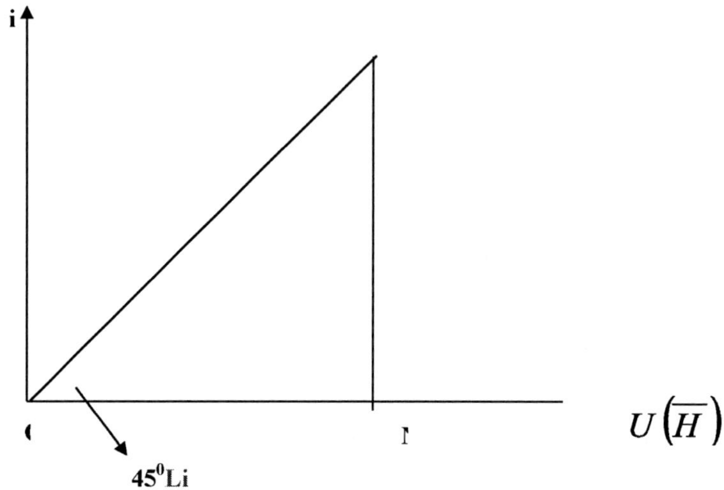

Fig. 12.2. Determination of the amount of child labour

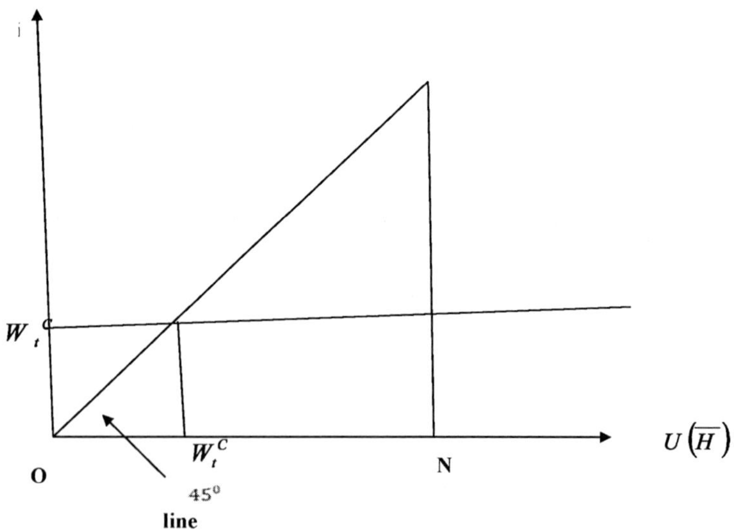

There are, as we have already mentioned, two segments in the production sector: an import-competing sector and an export sector. Both use only labour for production. The former employs both skilled and unskilled labour, and the latter employs only unskilled labour and child labour. It is a small open economy, with a tariff rate τ imposed on imports. World prices of exports and imports in terms of domestic currency are P_X and P_Y respectively. All the markets are perfectly competitive. Let us focus on the export sector first.

Export sector

This sector employs only adult unskilled labour and child labour for production. It may be sensible to assume that child labour and adult unskilled labour are homogeneous. The only difference is that a child in a given period supplies less labour than an adult unskilled worker. We assume that an adult unskilled worker supplies inelastically 1 unit of labour in any given period and a child supplies β fraction of the labour supplied by an adult unskilled labour in any given period. Accordingly, in equilibrium

$$W^C = \beta W^U \qquad 0 < \beta < 1 \qquad \text{------------------------------(4)}$$

where W^U denotes the wage rate of adult unskilled labour.

We assume that the production function in the export sector is a fixed coefficient and only a_{lu} amount of labour is needed per unit of output of the export good. This labour may be supplied by either adult unskilled workers or child workers or both.

Given the assumption of perfect competition, we have in equilibrium

$$a_{lu}W^U = a_{lu}\frac{W^C}{\beta} = P_X \qquad \text{---------------------------------(5)}$$

From (5) it follows that

$$W^U = P_X / a_{lu} \qquad \text{---------------------------------(6)}$$

$$W^C = \beta.(P_X / a_{lu}) \text{ ----------------------------------(7)}$$

(6) and (7) determine the wage rates. From (7) it follows that W_c is the same in every period. So the number of child workers and that of adult unskilled workers are the same in every period. Both are equal to W_C.

Import competing sector

The import competing sector produces only one good, the output of which is denoted by *Y*. The good is produced with both skilled and unskilled labour. Each adult worker supplies inelastically 1 unit of labour in any given period. The production function is Cobb-Douglas, given by

$$Y = S^{\alpha} U^{1-\alpha} \qquad \text{-------------------------------------(8)}$$

where S and U denote the amounts of skilled and unskilled labour employed in production in the import competing sector. The domestic price of *Y* is given by $P_Y(1+\tau)$. Since the import competing sector is perfectly competitive, the following equation must hold in equilibrium

$$P_Y(1+\tau) = a_{lsy}W_s + a_{luy}W_u \qquad \text{------------------------------(9)}$$

where, a_{lsy} and a_{luy} denote the amounts of skilled and unskilled labour required, respectively, per unit of output of *Y*. Let us now derive the values of a_{lsy} and a_{luy} by solving the following optimizing exercise:

$$\min\left[W_s S + W_u U\right]$$

s.t.

$$\bar{Y} = 1 = S^{\alpha} U^{1-\alpha}$$

The above minimization exercise yields

$$a_{lsy} = \left(\frac{\alpha}{1-\alpha}\right)^{1-\alpha} \left(\frac{W_u}{W_s}\right)^{1-\alpha} \text{ and } a_{luy} = \left(\frac{1-\alpha}{\alpha}\right)^{\alpha} \left(\frac{W_s}{W_u}\right)^{\alpha} \text{ --10)}$$

Substituting (10) into (9), we get

$$P_Y(1+\tau)=\left(\frac{\alpha}{1-\alpha}\right)^{1-\alpha}\left(\frac{W_u}{W_s}\right)^{1-\alpha}W_s+\left(\frac{1-\alpha}{\alpha}\right)^{\alpha}\left(\frac{W_s}{W_u}\right)^{\alpha}W_u \text{ ----(11)}$$

Substituting (6) into (11), we can solve it for W_S. It is given by

$$W_S=\left[\left(\frac{\alpha}{1-\alpha}\right)^{1-\alpha}+\left(\frac{1-\alpha}{\alpha}\right)^{\alpha}\right]^{-\left(\frac{1}{\alpha}\right)}\times\frac{\left[P_y(1+\tau)^{(1/\alpha)}\right]}{P_x^{\left[(1-\alpha)/\alpha\right]}}\times a_{lu}^{[(1-\alpha)/\alpha]}$$

Or $W_s = W[P_y(1+\tau), P_x]$ ---------------(12)

where $\dfrac{\partial W}{\partial P_y}<0, \dfrac{\partial W}{\partial r}>0$ and $\dfrac{\partial W}{\partial \tau}>0$

Since skilled labour is used only in the import-competing sector, the full employment of skilled labour implies the following equation to hold:

$$N-W_c=a_{lsy}Y=\left(\frac{\alpha}{1-\alpha}\right)^{1-\alpha}\left(\frac{P_x}{a_{lu}}\right)^{1-\alpha}(W_s)^{\alpha-1}Y$$

Substituting (7) into the above equation and rearranging the terms, we have

$$Y=(N-\frac{\beta P_x}{a_{lu}})\left(\frac{1-\alpha}{\alpha}\right)^{1-\alpha}\left[\left(\frac{\alpha}{1-\alpha}\right)^{1-\alpha}+\left(\frac{1-\alpha}{\alpha}\right)^{\alpha}\right]^{(\alpha-1)/\alpha}\left[\frac{P_y(1-\tau)}{P_x}\right]^{(1-\alpha)/\alpha}a_{lu}^{(1-\alpha)/\alpha}$$

or $Y\equiv Y\left[P_x, P_y(1+\tau)\right]$------------------(13)

where, $\dfrac{\partial W}{\partial P_y}<0, \dfrac{\partial W}{\partial r}>0$ and $\dfrac{\partial W}{\partial \tau}>0$

The amount of unskilled labour employed in the import competing sector, denoted U_Y, is using (6), (12) and (13) accordingly.

$$U_Y = a_{luy}Y = \left(\frac{1-\alpha}{\alpha}\right)^{\alpha}\left[\frac{W\left(\frac{P_Y(1+\tau)}{P_X}\right)}{P_X / a_{lu}}\right]Y(P_X, P_Y(1+\tau)) \equiv U_Y\left(\underset{-}{P_X}, \underset{+}{P_Y(1+\tau)}\right)$$

-----(14)

The remaining supply of unskilled labour is employed in the export sector. Unskilled labour is supplied by both unskilled workers and child workers. The total supply of such labour in every period is $W_C + \beta W_C = (1+\beta)\beta(P_x / a_{lu})$. The supply of X in every period is therefore

$$X = \frac{1}{al_u}\left[\frac{(1+\beta)\beta(P_x / a_{lu})}{a_{lu}} - U_Y\left(\underset{-}{P_X}, \underset{+}{P_Y(1+\tau)}\right)\right] \equiv U_X\left(\underset{+}{P_X}, \underset{-}{P_Y(1+\tau)}\right)$$

-----(15)

Proposition 1

The key variable is P_X. The lower its value, the less the wage rate of child labour—see (7)— and therefore the less the incidence of child labour.

Proposition 2

An increase in tariff rate on imports will raise W_S—see (12)—and thereby increase inequality. It will also, as follows from (13) and (15),raise Y and lower X.

Proposition 2 may be explained as follows. It follows from (12) that W_S goes up as the tariff rate rises. This leads to the substitution of unskilled labour for skilled labour in the production of Y. Since the supply of skilled labour is independent of the tariff rate, the output of Y goes up. This raises the employment of unskilled labour in the production of Y. This, in turn, implies a fall in the output of X since the amount of unskilled labour available for the production of X goes down.

References

Baland, J. M. & J. A. Robinson. "Is Child Labour Inefficient?" *Journal of Political Economy* 108 (4) (2000): 663–679.

Baland, J. M. & C. Duprez, "Are Fair Trade Labels Effective Against Child Labour?" CEPR Discussion Paper 6259. 2007

Basu, A. K., Chau, N. H., & U. Grote. "Guaranteed Manufactured without Child Labour: the Economics of Consumer Boycotts, Social Labeling and Trade Sanctions." *Review of Development Economics* 10 (3) (2006): 466–491.

Basu, A. K. & N. H. Chau. "Exploitation of Child Labour and the Dynamics of Debt Bondage." *Journal of Economic Growth* 9 (2) (2004): 209–38.

Basu, K. "Child labour: Cause, Consequence, and Cure, with Remarks on International Labour Standards."*Journal of Economic Literature* 38 (1999): 1083–1119.

Basu, K. "A Note of Multiple General Equilibria with Child Labour." *Economic Letters* 74 (3) (2002): 301–8.

Basu, K. & P. H. Van. "The Economics of Child Labour." *American Economic Review* 88 (3) (1998): 412–27.

Dessy S. E. & S. Pallage. "Some Surprising Effects Of Better Law Enforcement Against Child Trafficking." *Journal of African Development* 1 (1) (2006).

Dinopoulos & Zhao. "Child Labour and Globalization." *Journal of Labour Economics* 25 (2007): 553–579.

Doepke, M., & F. Zilibotti. "The Macroeconomics of Child Labour Regulation." *American Economic Review* 95 (5) (2005): 1492–1524.

—. "International Labour Standards and thePolitical Economy of Child-labour Regulation." *Journal of the European Economic Association* 7 (2009): 508–518.

—. "Do International Labour Standards Contribute to the Persistence of the Child-Labour Problem?" *Journal of Economic Growth* 15 (1) (2010): 1–31

Edmonds, E. V. "Child Labour." In *Handbook of Development Economics*, Volume 4, eds. Schultz, T. P & J. Strauss. Amsterdam: North Holland, 2008.

Edmonds, E. V. & N. Pavcnik. "Child Labour in the Global Economy." *Journal of Economic Perspectives* 18 (1) (2005): 199–220.

— "The Effect of Trade Liberalization on Child Labour." *Journal of International Economics* 65 (2) (2005): 401–419.

—. "International Trade and Child Labour: Cross-Country Evidence." *Journal of International Economics* 68 (1) (2006): 115–140.

Fan, C. S. "Relative Wage, Child Labour and Human Capital." *Oxford Economic Papers* 56 (4) (2004a): 687–700.

—. "Child Labour and the Interaction between the Quantity and Quality of Children." *Southern Economic Journal* 71 (1) (2004b): 21–35.

Genicot, G. "Malnutrition and Child Labour." *Scandinavian Journal of Economics* 107 (1) (2005): 83–102.

Gupta, M. R. "Wage Determination of a Child Worker: A Theoretical Analysis." *Review of Development Economics* 4 (2) (2000): 219–28.

Hazan, M. & B. Berdugo. "Child Labour, Fertility, and Economic Growth." *Economic Journal* 112 (482) (2002): 810–828.

Jafarey, S. & Lahiri, S. "Will Trade Sanctions Reduce Child Labour? The Role of Credit Markets." *Journal of Development Economics* 68 (1) (2002): 137–156.

Kambhampati, U. S. & R. Rajan. "Economic Growth: A Panacea for Child Labour?" *World Development* 34 (3) (2006): 426–445.

Maskus, K. E. "Should Core Labour Standards be Imposed Through International Trade Policy?" Policy Research Working Paper no. 1817. World Bank, Washington DC, 1997.

Ranjan, P. "Credit Constraints and the Phenomenon of Child Labour." *Journal of Development Economics* 64 (1) (2001): 81–102.

CHAPTER THIRTEEN

INTERNATIONAL FRAGMENTATION IN THE PRESENCE OF THE ALTERNATIVE HEALTH SECTOR SCENARIO: A THEORETICAL ANALYSIS

TONMOY CHATTERJEE AND KAUSIK GUPTA

1. Introduction

In recent years the importance of the health sector as a potential engine of growth of a developing economy like India has been argued by many contemporary economists. The health sector has shown a growth of 9.3 % between 2000–2009, comparable to the sectoral growth of other emerging economies such as China and Brazil.[1]The total **value** of the sector was more than $38 billion, about 5.1% of GDP (Ernst & Young 2008). The Indian Healthcare market is estimated to reach US $77 billion by 2013 (Pricewaterhouse Cooper 2007). The healthcare industry accounted for 5.1% of India's GDP in 2006. The compound annual growth rate of the Indian healthcare sector was 16% during the 1990s (Ibid.), and it is expected to grow at a compound annual growth rate (CAGR) of 15% over the next fifteen years (Ernst & Young 2008). It is also expected to generate employment to nine million people in 2012 (Ibid.).

In the recent past the recession in 2008 and the economic slowdown since 2011, intensified by the Eurozone crisis and the slowdown in the US economy, have brought about negativity in world economic growth projections. A recent report released by the United Nations (UN) shows that all developing economies will be affected by the slowdown. However, the good news is that the East Asian and South Asian economies are

---

[1] Yes Bank and Assocham report, as quoted by IBEF report on Healthcare accessed on January 30, 2012.

increasingly being seen as growth drivers of the world, following which the health sector has grown exponentially. A CII- Mckinsey report states that the Indian health sector has emerged as one of the largest service sectors with an estimated revenue of around $30 billion, constituting 5% of GDP and offering employment to around four million people. By 2025, the Indian population will reach 1.4 billion, with about 45% constituting urban adults.[2] To cater to this demographic change, the health sector will have to be about $100 billion in size, contributing 8–10% of future GDP. The growth in the health service sector would be driven by health care facilities, medical diagnostic and pathological laboratories and the medical equipment sector. All these growth inducing factors are related to medical equipment and devices. The medical equipment and devices industry is another sector closely tied up with the health service sector. In fact, the health service sector uses medical equipment and devices as an intermediate product for production. Advancement and innovations in medical technology have resulted in huge improvements in this intermediate sector. In India, over 65% of medical equipment is still being imported from abroad in a fast-growing domestic market, accounting for 80%–90% in the pre liberalization period.[3] It is often argued that developing countries like India import medical equipment and other related devices for its health service. Such a health service sector can be considered an exportable sector as the medical equipment and devices market is worth US $1,505 million in India, with the demand growing at about 15% per year.

It is to be noted that given the growing demand, the emergence of reputed private players and the huge investment needs in the healthcare sector have resulted in growing interest among the foreign players and non-resident Indians in entering the Indian healthcare market. The growing presence of corporate players and foreign investors in India's health sector, although highlighted and documented in various reports by different sources, are not yet well understood in terms of their current status as well as their implications for the health system at large. For example, while the emergence of corporate hospitals or foreign funding and tie ups in the hospital segment can have many positive implications, such as helping to improve the physical infrastructure, standards, quality of healthcare, technology and processes along with spillover benefits in areas such as medical devices, pharmaceuticals, outsourcing, and research and development may also result in higher costs of health care and greater

[2] *The Times of India*, February 2, 2012.
[3] This implies that the health service sector is internationally fragmented.

segmentation between the public and private health sectors. Thus, there is a need to examine whether there exists any constraint related to the inflow of foreign capital in the health sector. However, there are external and domestic factors which constrain foreign investment, especially foreign direct investment (FDI hereafter), in India's hospital segment. One of the external factors which has been noted is that, notwithstanding trends towards privatization in health in major developed countries, this is a sector that is undergoing reform and internal problems in those economies. In many countries, the number of private players who can establish hospitals overseas is limited. Hence, the potential number of overseas institutions that can invest in emerging markets may also be rather limited. Again, for some countries it is primarily domestic factors specific to the hospital business that are responsible for limiting the extent of FDI in India's hospitals, such as the initial establishment related factors as well as post-establishment related operational issues which affect the returns of investment. To remove these problems the state have taken some steps in the post-globalization period:

(1) reduction of import duty for medical equipment and devices to 5% with countervailing duty (CVD) of 4%, of which more than 50% is imported. Assistive devices, rehabilitation aids, etc. have been completely exempted from CVD (Union Budget 2010–2011).
(2) relaxed rules for the NRI medical practitioners to invest and provide medical services in India (Baru 1998) and depreciation rates for essential equipment and consumables increased from 25% to 40%, giving a tax-saving incentive to the healthcare institutions.
(3) introduction of a medical visa and a medical attendant visa for a period of one year with four multiple entries in 2005, and allowing 100% FDI in the healthcare sector in 2000.
(4) making long-term loans and capital cheaper for healthcare institutions due to the 2002–2003 and 2003–2004 union budget and introduction of 100% income tax exemption for a period of five years, for new hospitals with more than 100 beds located outside of the eight agglomerations (Finance Act 2008).

As a result of these policies, and due to huge demand, the health service sector provides incentives to the investors for investments from domestic as well as financial investors and private equity firms. Funds such as ICICI Ventures, IFC, Ashmore and Apax Partners invested about US $450 million in the first six months of 2008–2009, compared to US $125 million during the same period of the previous year. Feedback

Ventures expects private equity funds to invest at least US $1 billion during 2009–2013. Twelve percent of the US $77 million venture capital investments in July to September 2009 were in the healthcare sector. GE plans to invest over US $3 billion on R&D, US $2 billion to drive healthcare information technology and health in rural and underserved areas, and US $1 billion in partnerships, content and services, over the next six years. International clinic chain Asklepios International plans to invest US $100–200 million in the Indian healthcare market. Gulf-based group Dr Moopen is planning to invest US $200 million for setting up hospitals and eye-care centres across India. Meanwhile, Fortis is planning to invest US $55 million to expand its pan-India operations.

In the last decade, the medical devices and equipment industry has successfully attracted foreign direct investment though this sector, which was previously importing 50%–60%. From merely US $2.3 million in 2000 it reached US $147.69 million in 2009. Some of big foreign firms in the sector invested in India either directly or through collaborations and joint ventures, including GE (USA), Isoft (Australia), Proton Healthcare (USA) and Seimens (Germany) etc.

The term international fragmentation has been used widely in trade literature, and the notable contributors are Jones & Kierzkowski (2003), Deardoff (2001), Jones & Marjit (2001) and Marjit (2007; 2009). All the authors mentioned above have discussed either the causes behind the term or relate it with the pattern of trade. Maiti & Marjit (2007) have considered a partial equilibrium framework and have shown that international trade enhances the possibilities of fragmentation in the production process in the Indian context. Jones & Marjit (2008) have considered a general equilibrium framework and from which we can argue that trade may lead to more fragmented activities relative to autarky, even if one observes specialization. Again Marjit, Beladi & Chakraborty (2004) have shown that reduction in the price of intermediate product may lead to a zone which is more fragmented. They have also examined the impact of fragmentation on the skilled-unskilled wage gap.

Though there exists a large number of theoretical works related to international fragmentation, very few of them have used the general equilibrium structure. Unfortunately, at the theoretical level almost no work in a general equilibrium structure has been done to relate international fragmentation with the health sector. The present chapter attempts to fill up the lacuna in this line.

The present chapter is an extension of Marjit, Beladi & Chakraborty (2004), as in this model an exportable sector, a health service sector and a health intermediate sector have been introduced. The chapter attempts to examine not only the reasons behind international fragmentation but also the impact of such international fragmentation on the output levels of the health sector. The impact of FDI on the health sector in the presence of international fragmentation is also of concern.

In this chapter we examine: (i) the impact of movement towards international fragmentation of the health sector in the absence of trade liberalization, and (ii) impact of trade liberalization on the health service sector in the presence of international fragmentation. To do so we have considered two different models. In model 1 we have shown the impact of movement towards international fragmentation of the health sector in the absence of trade liberalization. As the economies liberalize, competition in all the markets should increase. To capture the impact of such liberalization on the output level of the health service sector in the presence of international fragmentation we have also considered model 2.

This chapter is organized in the following manner. Section 2 considers model 1, as well as the drive towards fragmentation and the health sector. Section 3 considers model 2 and is divided into three subsections: 3.1 considers FDI in the health sector, 3.2 considers international health capital immobility and 3.3 considers international health capital mobility. Finally, the concluding remarks are made in section 4.

2. Model 1

We consider a small open economy consisting of four sectors in a Heckscher-Ohlin-Samuelson framework. Out of the four sectors, one is an agricultural sector(A) which produces an exportable good (X_A) using unskilled labour(L) and capital(K). The second sector is a manufacturing sector(M) which produces importable goods (X_M) using skilled labour (S) and capital. K is perfectly mobile between sectors A and M. The third and fourth sectors of our economy are the domestic intermediate health goods producing sector (I) and the health sector (H), respectively. Sector I uses skilled labour along with health capital (N) for production of the intermediate health product (X_I) of our economy and the health sector uses health capital,[4]skilled labour and intermediate health input (X_I) to produce another exportable product (X_H). Here we assume that the requirement of

[4] By the term health intermediate goods we actually mean those commodities which are exhausted due to the course of production in the health service sector(H), e.g. injectable goods and theirassociated products, several chemicals, equipment used in pathology and different forms of medicines. Again, by health capital we mean theequipment and products which are not exhausted due to the production process, e.g. ECG machine, X-ray machine etc.

intermediate goods for the production of one unit of output of the health sector is fixed. Unskilled labour (L) has been considered as specific to the agricultural sector (A). Skilled labour is perfectly mobile among sectors M, H and I. The skilled wage rate in the health sector is assumed to be fixed at a higher level compared to the skilled wage rate prevailing in the rest of the sectors. Here, health capital is perfectly mobile between sectors I and H. Health capital consists of both domestic health capital (N_D) and foreign health capital(N_F), and additionally we assume that N_D and N_F are perfect substitutes. This implies that an increase in foreign health capital will lead to an increase in the overall health capital endowment of the economy.

The agricultural product is considered as the numeraire and its price is set equal to unity. We assume that both foreign capital income and foreign health capital income are fully repatriated. Markets are competitive. Production functions in each sector exhibit constant returns to scale with diminishing marginal productivity to each factor. The following notations are used in this model:

X_i=product produced by the ith sector, i=A, M, I, H
P^*_A=world price of commodity A
P_A=domestic price of commodity A, we assume $P_A=P^*_A=1$
P_M=world price of good M
P_I=domestically determined price of good I
P^*_I=price of the foreign intermediate commodity
P_H=world price of good H
L=fixed number of unskilled workers in the economy
S=stock of skilled labour
N_D=domestic health capital stock of the economy
N_F=foreign health capital stock of the economy
N=economy's aggregate health capital stock ($N = N_D + N_F$)
K=economy's aggregate capital stock
a_{ji}=quantity of the jth factor for producing one unit of output in the ith sector, j=L,S,K,N and i=A,M,I,H
θ_{ji}=distributive share of the jth input in the ith sector
λ_{ji}=proportion of the jth factor used in the production of the ith sector
W=competitive unskilled wage rate
W_S=skilled wage rate
r=rate of return to capital
R=rate of return to health capital
σ_i=elasticity of factor substitution in sector i, i=A, M, I, H.
^=proportional change

The equational structure of the model is as follows.

The competitive equilibrium conditions in the product market for the four sectors give us the following equations.

$a_{LA}W + a_{KA}r = 1 (1)$
$a_{SM}W_S + a_{KM}r = P_M (2)$
$a_{SI}W_S + a_{NI}R = P_I (3)$
$a_{SH}\overline{W}_S + a_{NH}R + a_{IH}P_I = P_H (4)$

For simplicity we assume that a_{SH} and a_{IH} are given to us.

Equilibrium condition for the health intermediate sector is given by
$a_{IH}X_H = X_I (5)$

Sector specificity of unskilled labour is given by the following equation

$a_{LA}X_A = L (6)$

Perfect mobility of capital between sectors A and M can be expressed as

$a_{KA}X_A + a_{KM}X_M = K_D + K_F = K (7)$
Full employment of skilled labour implies the following equation
$a_{SM}X_M + a_{SI}X_I + a_{SH}X_H = S (8)$
Perfect mobility of health capital between sectors H and I can be expressed as
$a_{NH}X_H + a_{NI}X_I = N_D + N_F = N (9)$

The working of the model is as follows. In this model we have nine equations with nine endogenous variables, namely W, W_S, r, R, P_I, X_A, X_M, X_I, X_H, that is, the system is solvable. From equation (1) we can express W as a function of r. Similarly, from equation (2) we can express W_S in terms of r. From equation (4) we find that R is a function of P_I. Using this fact in equation (3) we can express W_S in terms of P_I. Hence, from equations (1) and (2) we can express W and r as a function of P_I only. In this model we cannot determine factor prices independently from the competitive equilibrium conditions. Thus, the structure is an indecomposable structure.[5] Using equations (6), (7), (8) and (9) we can

[5] If the factor prices are determined independently of factor endowments we refer to the structure as a decomposable structure.

express X_A, X_M, X_I and X_H as a function of P_I. Thus, from equation (5) we can determine the value of P_I, as X_H and X_I are function of P_I. Once P_I is known W, W_S, r, R, X_A, X_M, X_I, X_H are also known.

In this section we want to analyze several causes due to which fragmentation is possible. In this model initially we considered sector I as a domestic intermediate good producing sector. This intermediate good can be imported by the health sector from overseas but in this case it has to incur a fixed cost (F) mainly due to transaction or communication factors.[6] This is the main constraint for the health sector in buying foreign intermediate goods and leads to an increase in demand for domestic intermediate goods. Given supply, an increase in demand for I leads to an increase in P_I. It implies that no fragmentation situation is associated with a higher level of P_I. Here, we are starting with a situation where domestic intermediate product price is greater than international intermediate product price, i.e. $P_I > P^*_I$. From equation (4) we can say that a fall in P_I leads to an increase in R, since P_H and W_S are exogenously given.[7] Let R^* be the rate of return on health capital corresponding to the price of the intermediate good P^*_I. It is to be noted that P^*_I and R^* are also negatively related by the similar argument as we use in the case of P_I with R. Let R^0 be the initial equilibrium level of R. Then we can say $(R^* - R^0)$ also varies inversely with P_I^*. This is because R^* changes due to a change in P_I^*, for given R^0. So we can say that the health sector uses domestic intermediate goods; that is, fragmentation is not preferable, iff $(R^* - R^0)\,N < F$.

Using the similar argument we can consider different cases, that is:

case 1: fragmentation is preferable iff $(R^* - R^0) > F/N$
case 2: fragmentation is not preferable iff $(R^* - R^0) < F/N$;
case 3: health sector will be indifferent iff $(R^* - R^0) = F/N$.

The above analysis can be explained with the help of Fig. 12.1.

---

[6] For details see Marjit, Beladi & Chakraborty (2004).
[7] For details see Appendix A.

Figure 12.1

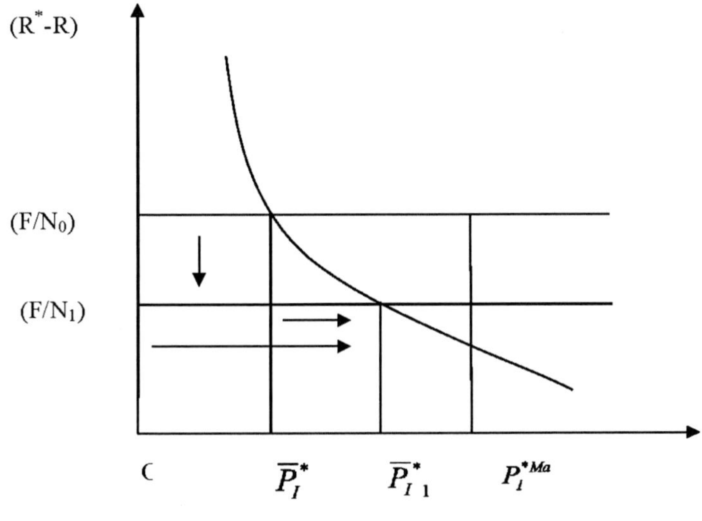

In this chapter we find that there exist two critical values of P_I^* such that for all $P_I^* \in [\overline{P}_I^*, P_I^{*max}]$ there is no fragmentation, where $\overline{P}_I^*$ is the lower critical value of P_I^* and P_I^{*max} is the upper critical value of P_I^*. Here P_I^{*max} is nothing but the domestic market determined value of P_I. It is to be noted for $P_I^* \in (0, \overline{P}_I^*)$ that there will be fragmentation. Thus, it is clear from the above figure that $\overline{P}_I^*$ is the lower critical value of P_I^* and it is the maximum price of the foreign intermediate for which the health sector will go for fragmentation. In this figure the area left of $\overline{P}_I^*$, that is, where $\overline{P}_I^* > P_I^*$, we have a situation of fragmentation whereas the area to the right of $\overline{P}_I^*$ up to P_I^{*max} gives us a situation where fragmentation is not possible. For simplicity here we consider $P_I^* \in [\overline{P}_I^*, P_I^{*max}]$ as the relevant interval for P_I^*.

From case 2 we can infer that fragmentation may be possible due to either a fall in F or an increase in N (or N_F hereafter). It is to be noted that a fall in F or an increase in N_F leads to a downward shift of the F/N schedule. It implies an expansion of the area to the left of $\overline{P}_I^*$ and contraction of the area of its counterpart. Again, for given F and N, fall in P_I towards the lower limit of P_I^* ($\overline{P}_I^*$) leads to a situation where the possibility of fragmentation increases. We state the results in the form of following proposition:

Proposition 1: *A fall in the price of domestic health intermediate goods, reduction in fixed cost or an inflow of foreign health capital enhance the possibility of international fragmentation in the health market.*

2.1 The drive towards fragmentation and the health sector

So far we have analyzed the causes and possibilities of fragmentation. In this section we are trying to focus on the drive towards fragmentation in the presence of the health sector. This is captured through a fall in domestic price (P_I) of the intermediate product.[8] We have already mentioned that fall in P_I towards $\overline{P}_I^*$ implies a shift from a regime of no fragmentation to a regime of fragmentation.[9] Here, we have to examine the intermediary effects of such a regime change.

From equation (4) we can argue that a fall in P_I leading to an increase in R, as P_H and W_S are given. From equation (3) we can say that a reduction in P_I and an increase in R lead to a situation where to maintain equality W_S must have to fall. Again, reduction in W_S leads to an increase in r, as P_M is given in equation (2). Using this fact in equation (1) we get a fall in W. This fall in W implies an increase in a_{LA} so that to maintain unskilled labour market equilibrium condition X_A must go down. Again, a fall in P_I leads to an increase in r and hence a_{KA}, a_{KM} must go down. Thus from equation (7) we can get an increase in X_M. Using equation (5) in equation (9) we obtain

$$(a_{NH}+a_{NI}a_{IH})X_H=N \quad (9')$$

From equation (9′) we can easily argue that X_H must go up due to a fall in P_I [as a_{NH} and a_{NI} decrease due to an increase in R]. Similarly, equation (8) can be written as

---

[8] Here we assume that fragmentation is possible only through reduction P_I because the change of F and N are significantly low in the regime where trade liberalization is absent.

[9] Here we consider a finite change of P_I.

$a_{SI}X_I=(S-a_{SH}X_H-a_{SM}X_M)(8')$

Fall in P_I leads to a fall in W_S. Reduction in W_S leads to an increase in the levels of both a_{SM} and a_{SI}. Thus, increase in the levels of a_{SH} X_H, a_{SM} X_M and a_{SI} lead to reduction of the right-hand side and expansion of the left-hand side of equation (8'). Hence, to maintain skilled labour market equilibrium condition X_I must go down.[10] Thus, the following proposition can now be established.

Proposition 2: *A movement from a regime of no fragmentation towards a regime of fragmentation leads to an increase in the levels of output of both the health and manufacturing sectors and a reduction in the level of output of the agricultural sector.*

3. Model 2

3.1 FDI in the Health Sector

In the earlier model we have analysed the impact of movement from a no fragmentation regime to a regime of fragmentation on the output levels of the health sector. In this model we are trying to analyze the impact of liberalization on the health sector in the presence of fragmentation in the new set up. In this model we have assumed that the skilled worker of the health service sector is also earning a competitive skilled wage.[11] In this chapter we consider that the total foreign health capital stock consists of both domestic health capital and foreign health capital. We have considered two regimes here. One is the regime of international health capital immobility and the second of international health capital mobility. In the context of the first regime we have considered foreign health capital as exogenous, implying the existence of foreign health capital immobility. In the second regime we have considered foreign health capital as endogenous.

---

[10] Here we have implicitly assumed that the supply of an intermediate health product is the sum of X_I and X_I^+, where X_I^+ is the amount of import of intermediate health product. Thus,theactual equilibrium condition for the health intermediate market is a_{IH} $X_H = X_I + X_I^+$. But in the absence of fragmentation it implies $X_I^+ = 0$. That is why we have considered equation (5) initially. It is to be noted that as we move towards international fragmentation we get an increase in X_H and a reduction of X_I. Thus,tomaintainthe above equality X_I^+ must go up as an adjustment term.

[11] Here, we also assume that a_{SH} is a fixed input-output coefficient, as we assume in model 1.

3.2 International health capital immobility

We consider a small open economy where international health capital is immobile[12] and consists of four sectors in a Heckscher-Ohlin-Samuelson framework. The modified equational structure can be written as follows.

The competitive equilibrium conditions in the product market for the three sectors give us the following equations:

$a_{LA}W + a_{KA}r = 1$(1)

$a_{SM}W_S + a_{KM}r = P_M$(2)

$a_{SH}W_S + a_{NH}R + a_{IH}\overline{P}_I^* = P_H$ (3)

For simplicity we assume that a_{SH} and a_{IH} are given to us.

The equilibrium condition for the intermediate health product is

$a_{IH}X_H = X_I^+$(4)

Here, X_I^+ is the amount of import of intermediate products.

Sector specificity of unskilled labour is given by the following equation

$a_{LA}X_A = L$(5)

The perfect mobility of capital between sectors A and M can be expressed as $a_{KA}X_A + a_{KM}X_M = K_D + K_F = K$(6)

Full employment of skilled labour implies the following equation

$a_{SM}X_M + a_{SH}X_H = S$(7)

Sector specificity of foreign health capital is given by the following equation

$a_{NH}X_H = N_D + N_F = N$ (8)

---

[12] International health capital immobility is a situation where the domestic rate of return on foreign health capital (R) is greater than the rate of return on foreign health capital in the international market (R^+), and there is restriction on the entry of foreign health capital to the domestic economy.

In this model we have eight equations with eight unknowns, namely W, W_S, r, R, X_A, X_M, X_I^+, X_H; that is, the system is solvable. From equations (1), (2) and (3) we can express W, W_S and R as a function of r. In this model we cannot determine factor prices independently from the competitive equilibrium conditions. Thus, the structure is an indecomposable structure. Using (5) and (6) we can express X_A, X_M in terms of r. Thus, from equation (7) we can express X_H as a function of r. Again, from equation (8) one can express X_H in terms of r, as N is given. Hence, by using equations (7) and (8) we can determine the values of X_H and r. Once r is known W, W_S, R, X_A, X_M, X_I^+ are also known.

3.3 International health capital mobility

Here, we assume that in the presence of international health capital immobility we have $R > R^+$, where $R^+$is the given return on foreign health capital in the international market. In such a situation we have no foreign health capital inflow. If R falls to $\tilde{R}$, where, $R>\tilde{R}> R^+$, we find that there is some amount of inflow of foreign health capital (N_F) and we will at last reach the equilibrium level[13] of N_F where, $R = R^+$.

Here, we assume that N_D is exogenous, whereas N_F is assumed to be an endogenous variable and we use $R = R^+$ in model 2. Thus, here we also have eight independent equations with eight endogenous variables, so the system can be solved. Using $R= R^+$ from equations (2) and (3) we can solve for W_S and r (as $P_I= \overline{P}_I^*$). Once W_S and R are known a_{NH} is also known. Thus, from equation (8) we can express X_H as a function of N_F, as R is already known and W_S can be explained in terms of R. Using equations (5) and (6) we can determine the values of X_A and X_M. Hence, from equation (7) we can derive the value of X_H. Once X_H is known then N_F and X_I^+ can also be determined from equation (8) and (4) respectively, and this completes the working of the model.

Explanations of sectoral effects due to liberalization are given below. An increase in N_F implies a fall in R. From equation (3) we can say that a fall in R leads to an increase in W_S. From equations (1) and (2) we can argue that an increase in W_S implies an increase in W and a reduction in r.[14] Since an increase in W implies a reduction in a_{LA}, maintaining unskilled labour market equilibrium condition X_A must go up. Again, a fall in r leads to an increase in a_{KA} and a_{KM}. Thus, from equation (6) we can

---

[13] At $R=R^+$, we have the equilibrium level of foreign health capital inflow due to equilibrium in the International health capital market.

[14] See Appendix B.

get a reduction in the output level of sector M (X_M). A fall in R due to an inflow in N_F leads to an increase in W_S. An increase in W_S leads to a fall in a_{SM}. Thus, a fall in a_{SM} X_M leads to an increase in (S – a_{SM} X_M), that is, the availability of skilled labour increases to the health sector and induces it to expand.[15] From equation (5) we can argue that X_I^+ must go up as X_H increases due to international health capital mobility.

Proposition 3: *A shift from a regime of international health capital immobility to health capital mobility causes an increase in the levels of output of both the health and agricultural sectors and a reduction in the level of output of the manufacturing sector.*

4. Concluding Remarks

The health service and health intermediate sector (medical devices and equipment sector) are gaining more importance among economists, and hence these sectors become the most important parts of the social sector of any developing economy like India. The present chapter has considered two different models. In the first model we build up a structure (based on the H-O-S general equilibrium structure) where we introduce the health service sector as an export sector, and more interestingly we assume that the production function of that sector can be fragmented. Here, we also assume that the wage of skilled labour in the health service sector is fixed at a higher level compared to the competitive skilled wage rate. The above-mentioned sectors use a special type of capital (health capital). In such a set-up we have shown that an increase in foreign health capital or a decrease in the price of health intermediate products may increase the possibilities of fragmentation. Apart from this, from that model we have also shown that a change of regime from no fragmentation towards fragmentation leads to an increase in the levels of output of both health and manufacturing sectors and a reduction in the level of output of the agricultural sector.

In the second model we have considered a three-sector general equilibrium structure where the third sector is a health service sector and its production process is fragmented. Here we have examined the impact of trade liberalization in the form of regime change on the output levels of different sectors in the presence of fragmentation. In this part we have shown that a change in regime from international health capital immobility

[15] As a_{SH} is given, from equation (7) we can show that an increase in (S- $a_{SM}X_M$) must increase in X_H, otherwise the skilled labour market will be in disequilibrium.

to international health capital mobility leads to the expansion of both health sector and agricultural sector and contraction of manufacturing sector.

References

Deardorff, A. V. "Fragmentation in Simple Trade Models." *The North American Journal of Economics and Finance* 12 (2) (2001): 121–137.

Jones, R. W. & Kierzkowski, H. "International Fragmentation and the New Economic Geography." *The North American Journal of Economics and Finance* 16 (1) (2005): 1–10.

Jones, R. W. & Marjit, S. "The Role of International Fragmentation in the Development Process." *American Economic Review* 91 (2) (2001): 363–366.

Jones, R. W. & Marjit, S. "Competitive Trade Models with Real World Features." *Economic Theory* 2008.

Lahiri, K. "FDI in India's Healthcare Sector: an Assessment." Paper Presented at Annual Seminer, Rabindra Bharati University, Kolkata, 2012.

Marjit, S. "Trade Theory and the Role of Time Zones." *International Review of Economics and Finance* 16 (2) (2007): 153–160.

Marjit, S. "Two Elementary Propositions on Fragmentation and Outsourcing in Pure Theory of International Trade." *Mimieo*, CSSS, Calcutta. (2009):

Marjit, S. & Gupta, K. "International Capital Mobility and Child Labour." Paper presented at the 4th Annual Conference on Economic Growth and Development, *Indian Statistical Institute*, New Delhi, 2008.

Marjit, S., Beladi, H. & Chakraborty, A. "Trade and Wage Inequality in Developing Countries." *Economic Enquiry* 42 (2): (2004): 295–303.

Marjit, S. & Kar, S. "Emigration and Wage Inequality." *Economics Letters* 88 (2005): 141–145.

—. "Emigration, Wage Inequality and Vanishing Sector." *MPRA* No.19354, 2009.

Appendix A

Differentiation of equation (4) gives us

$$\theta_{NH}\,\hat{R} + \theta_{NH}\,\hat{a}_{NH} + \theta_{IH}\,\hat{P}_I = 0$$

Nowσ_H= $(\hat{a}_{NH} - \hat{a}_{IH} / \hat{P}_I - \hat{R})$

$$\hat{a}_{NH} = (\hat{P}_I - \hat{R})\sigma_H$$

Using it we get

$$\theta_{NH}(1 - \sigma_H)\,\hat{R} + (\theta_{NH}\,\sigma_H + \theta_{IH})\,\hat{P}_I = 0$$

$$\hat{R} = (A_2/A_1)\,\hat{P}_I \quad (A.1)$$

Where, A_1, $A_2 > 0$.
Differentiation of equation (3) gives us

$$a_{SI}\,dW_S + W_S\,da_{SI} + a_{NI}\,dR + R\,da_{NI} = dP_I$$

From the envelop condition we get

$$W_S\,da_{SI} + R\,da_{NI} = 0$$

Using envelop condition we get

$$\theta_{SI}\,\hat{W}_S + \theta_{NI}\,\hat{R} = \hat{P}_I$$

Using (A.1) we get

$$\hat{W}_S = \{(1/\theta_{SI}) + (\theta_{NI}/\theta_{SI})(A_2/A_1)\}\,\hat{P}_I$$

$$\hat{W}_S = A_3\,\hat{P}_I \quad (A.2)$$

Where $A_3 > 0$.

Differentiating equation (2) we get

$\hat{r}$ =-$(\theta_{SM}/\theta_{KM})\hat{W}_S$ (A.3)

Similarly from equation (1) one obtains

$\hat{W}$=-$(\theta_{KA}/\theta_{LA})\hat{r}$ (A.4)

Differentiation of (7) gives us

$\hat{X}_H=(1/\lambda_{IH})\hat{X}_I$ (A.5)

Differentiating equation (6) we get

$\lambda_{NH}\hat{X}_H+\lambda_{NH}\hat{a}_{NH}+\lambda_{NI}\hat{X}_I+\lambda_{NI}\hat{a}_{NI}=0$(6.A)

We know,

$\sigma_I=(\hat{a}_{NI}-\hat{a}_{SI}/\hat{W}_S-\hat{R})$(A.6)

Again $W_S\, da_{SI} + R\, da_{NI} = 0$

$\hat{a}_{SI}$=-$(\theta_{NI}/\theta_{SI})\hat{a}_{NI}$(A.7)

Using (A.7) in (A.6) we get

$\hat{a}_{NI}=\sigma_I\theta_{SI}(\hat{W}_S-\hat{R})$(A.8)

Inserting the value of (A.8) in equation (6.A) and simplifying we get

$(\lambda_{NH}+\lambda_{NI}\lambda_{IH})\ \hat{X}_H = (-\lambda_{NH}\sigma_H)\ \hat{P}_I + (-\lambda_{NI}\theta_{SI}\sigma_I A_3)\ \hat{P}_I + (\lambda_{NH}\sigma_H + \lambda_{NI}\theta_{SI}\sigma_I)(-A_2/A_1)\ \hat{P}_I$

$\hat{X}_H=(A_5+A_6+A_7/A_4)\hat{P}_I$ (A.9)

Where, A_5,A_6, A_7<0 and A_4> 0.

Differentiation of equation (6) gives us

$\lambda_{LA}\hat{a}_{LA}+\lambda_{LA}\hat{X}_A=0$(6.A)

We know, $\sigma_A = (\hat{a}_{KA}-\hat{a}_{LA}/\hat{W}-\hat{r})$

$\hat{a}_{LA}=\hat{a}_{KA}-\sigma_A(\hat{W}-\hat{r})$(A.10)

Using envelop condition we get

$\hat{a}_{KA}=-(\theta_{LA}/\theta_{KA})\hat{a}_{LA}$(A.11)

Using (A.11) in equation (A.10) and simplifying we get

$\hat{a}_{LA}=-\sigma_A(\theta_{KA}\theta_{SM}/\theta_{KM}\theta_{LA})A_3\hat{P}_I$ (A.12)

Using (A.12) in (6.A) we can get

$\hat{X}_A=\sigma_A(\theta_{KA}\theta_{SM}/\theta_{KM}\theta_{LA})A_3\hat{P}_I$ (A.13)

Differentiation of equation (7) gives us

$\lambda_{KM}\hat{X}_M=\lambda_{KA}\hat{a}_{KA}-\lambda_{KM}\hat{a}_{KM}-\lambda_{KA}\hat{X}_A$

Using (A.11) and (A.12) we get

$\hat{a}_{KA}=A_8\hat{P}_I$ (A.14)

Where, $A_8=\sigma_A(\theta_{KA}\theta_{SM}/\theta_{KM}\theta_{LA})A_3(\theta_{LA}/\theta_{KA})>0$.

Using the elasticity of substitution of sector M and envelop condition we can get

$\hat{a}_{KM}=\sigma_M(\theta_{SM}/\theta_{KM})A_3\hat{P}_I$ (A.15)

Inserting the values of (A.13), (A.14) and (A.15) one can obtain

$\hat{X}_M = [\{\lambda_{KA}\ A_8 - \lambda_{KM}\sigma_M\ (\theta_{SM}/\theta_{KM})\ A_3 - \lambda_{KA}\sigma_A\ (\theta_{KA}\theta_{SM}/\theta_{KM}\theta_{LA})\ A_3\}/\lambda_{KM}]\ \hat{P}_I$ (A.16)

Appendix B

Differentiating equation (3) and using envelop condition we get

$$\theta_{SH}\hat{W}_S + \theta_{NH}\hat{R} = 0$$

$$\hat{W}_S = -(\frac{\theta_{NH}}{\theta_{SH}})\hat{R} \text{ (1.B)}$$

Differentiating equation (2) we get

$$\hat{r} = -(\theta_{SM}/\theta_{KM})\hat{W}_S \text{ (B.1)}$$

Using (1.B) in equation (B.1) we get

$$\hat{r} = (\theta_{SM}\theta_{NH}/\theta_{KM}\theta_{SH})\hat{R}$$

$$\hat{r} = B_1\hat{R} \text{ (2.B)}$$

Where, $B_1 > 0$.

From equation (1) after differentiation one obtains

$$\hat{W} = -(\theta_{KA}/\theta_{LA})\hat{r} \text{ (B.2)}$$

Using (2.B) in equation (B.2) we can obtain

$$\hat{W} = -(\theta_{KA}/\theta_{LA})B_1\hat{R} \text{ (3.B)}$$

SECTION D:

NATURAL RESOURCES AND LIVELIHOOD

CHAPTER FOURTEEN

POTENTIAL OF ORGANIC FARMING FOR PROVIDING SUSTAINABLE LIVELIHOOD: A STUDY IN EAST SIKKIM

RUMA KUNDU

1. Introduction

As a technique of cultivation, organic farming is still in the process of gaining popularity in India. The technique, which mainly evolved in the developed countries of the West, has been looked upon as an alternative to conventional methods of agriculture.

Organic farming utilises techniques like crop rotation, green manure and compost, biological pest control and mulching. An important characteristic is that it involves the use of fertilisers and pesticides while excluding or strictly limiting the use of manufactured or synthetic fertilisers and pesticides.

One of the major reasons for the rising popularity of organic agriculture is that it has several advantages over the conventional agricultural techniques, preserving the health of soil, ecosystems and people. It utilises ecological processes, biodiversity and cycles adapted to local conditions, instead of using inputs with adverse effects.

As far as India is concerned, the state of Sikkim can be undoubtedly regarded as a pioneer in organic agriculture, though the organic movement is also taking root in other states. Sikkim is naturally endowed with suitable climatic conditions that enable the state to organically cultivate crops with high demand in both domestic and international markets, like cardamom, ginger, pulses, turmeric, buckwheat, baby corn, sweet corn and medicinal herbs. As of 2010, 42,000 hectares (about 60% of the total cultivable land) were used for organic farming. A strong road map has been devised by government agencies under which the sale of chemical fertilisers and pesticides was completely banned and subsidies on chemical fertilisers were completely withdrawn in the state in May 2003. Simultaneously, the large-scale use of bio-fertilisers has also been widely promoted.

While this is the general cause behind livelihood diversification, in Sikkim the situation is different. Here, a peculiar problem is being witnessed. Even though organic farming as an agricultural practice is being supported from the side of the government both at the policy level and also in terms of resource allocation, a certain reluctance appears to be developing among the practitioners of this type of agriculture. This is because, unlike conventional agriculture, the expected benefits from organic farming are not obtained immediately. However, the farming community is not patient enough to wait for the desired outcome. Thus, due to the lack of immediate returns, those who started this mode of cultivation are moving on to other forms of livelihood. This is leading to livelihood diversification which, if continued, can harm agriculture in the state in the long run.

The present work is concerned with examining the potential of organic farming as a means for sustainable livelihood.

The remaining portion of the chapter consists of the following. Section II considers the brief literature survey and the justification for the study. Section III introduces the objectives of the study. Section IV explains the database and methodology for the study. Section V considers the results and discussions. Section VI gives the conclusive observations.

2. Brief literature survey and justification for the study

A considerable amount of research has taken place in the field of organic farming, though more at the international level. It would be relevant to consider some of the work done in this direction.

2.1 International Status

Rigby & Caceres (2001) have dealt with the issue of organic farming and sustainability of agricultural systems. The authors discuss the extent of operational meaning for sustainable agriculture. In this connection they address issues like the role of regulation and the use of synthetic agrochemicals, the desired degree of self-reliance of agricultural systems, and the scale of production and trade in agricultural goods.

Heaton (2001) has carried out a review of literature on organic agriculture on behalf of the Soil Association, Bristol, Great Britain, with a view to emphasising the myriad benefits of this type of farming activity.

The review touches on issues like food safety, the role of nutrients, the health effects of organic food, etc.

Greene & Kremen (2003) have described the status of organic farming in the USA during the period 2000–2001 *vis-a-vis* the adoption of certified systems. An initiative started during the 1990s, it has gathered momentum as the farmers try to keep up with consumer demand at both the local and national levels. The report updates USDA estimates of land farmed with organic practices during 1997 with estimates for 2000 and 2001, and provides new estimates on the number of certified organic operations in each state.

Mollá-Bauzá et al. (2005) contributed a study evaluating the premium that Spanish customers are willing to pay for organic wine in comparison to conventional wine with similar characteristics. For this purpose they employed the contingent valuation method involving a survey and direct estimation of premium price. The question pattern was based on dichotomous choice valuation with follow-up questioning. The methodology involved descriptive statistical analysis and logistic regression with the comparison of the two estimates. The evaluation of premium prices has been carried out for three segments of customers based on their lifestyle preferences. The authors conclude that customers with a preference for healthy lifestyles are prepared to pay a premium price for organic products.

The study carried out by Niggli et al. (2007) under the aegis of the International Trade Centre UNCTAD/WTO concentrates on organic agriculture and its adaptability to the effects of climate change, those both predictable and unpredictable. In doing so it throws light on the contribution of agriculture to climate change, the potential of organic agriculture for reducing the emission of greenhouse gases and its contribution towards carbon dioxide sequestration in the soil. It also considers the weaknesses of organic agriculture with regard to climate change and the inclusion of organic agriculture in voluntary carbon dioxide markets.

Rodriguez et al. (2008) have attempted to estimate customers' willingness to pay for organic products that are available in the domestic market of Argentina. Their objective was to gain insights into and propose strategies for the production, marketing, regulation and labelling of organic food products. The authors have applied the binomial multiple logistic regression model with the help of data collected from a food consumption survey carried out in Buenos Aires. They have considered the contingent valuation method for five kinds of products: regular milk, leafy vegetables, whole wheat flour, fresh chicken and aromatic herbs. On the basis of the results the authors opine that while customers are willing to pay a premium price for these products, there are constraints in the form of lack of storage capacity and a reliable regulatory system for reducing risks regarding quality.

The study by Chouichom & Yamao (2010) relates to a comparison of the opinions and attitudes of farmers involved in the cultivation of jasmine rice on the two sides of the organic divide. One hundred interviewees in each category were questioned as a part of the survey. The objective was to gauge their feelings on issues like organic farming knowledge, environment, marketing, and costs and benefits. Socio-economic indicators were also considered. Statistical tools utilised for this purpose included chi-square and t tests. The results indicate a correlation of attitudes in both categories with respect to the four aspects considered. It was also found that factors like educational level, farm holding and extension worker contract influenced the opinions of organic farmers. On the other hand, in the case of conventional farmers, their farming practices affected their opinions regarding organic farming.

Thus, it can be seen that the work on organic farming at the international level is much more mature and diverse, dealing with a variety of areas relating to the subject.

2.2 National Status

Narayanan (2005) has turned his attention towards the condition of organic farming in India. He contends that in spite of the conventional system of agriculture not yielding the desired benefits, India has lagged with regard to the application of the practice of organic farming compared to the rest of the world. In his opinion the relative success of other nations in this area can be traced to factors like high awareness of the health problems caused by the consumption of contaminated food products, the ill effects of environment degradation, and appropriate supports by the government and organisations like the European Union and International Federation of Organic Agriculture Movements (IFOAM). With conventional agriculture, it is difficult to keep up with the demands of a growing population in India and, together with a very negligible percentage of land under organic farming, the picture is not very bright. NGOs have played a major role in the growth of the organic farming movement in India. The author identifies existing weaknesses in the absence of linkages between farmers and markets and lack of financial support from the government.

Shiva (2005) upholds organic farming as the new revolution for agriculture in India, and discusses how organic farming can serve as the vehicle for the achievement of the Millennium Development Goals.

Mandal et al. (2008) carried out a study on the issue of biodiversity and organic farming in the north-eastern part of India, focussing on the opportunities and challenges involved. They have thrown light on how organic farming can contribute to enhancing biodiversity in this region, putting forward various suggestions in this regard such as the creation and maintenance of databanks, in-depth analysis for gaining an understanding of farmers' knowledge and expertise, selection/breeding and utilization and management of plant genetic resources, detailed understanding of the changes in the local species and varieties and those used by farmers, etc.

Venkareswarlu (2008) examines the prospects and limitations of organic farming with reference to rainfed agriculture. He puts forward a set of recommendations for realising the full potential of this new technique in rainfed agriculture.

Reddy (2010) provides a review of the status, issues and prospects in organic farming. The author covers the scenario for organic farming with regard to the world at large and India in particular. He also discusses the potential for organic farming in dryland regions.

Subrahmanyeswari & Chander (2011) put forward the unique proposal of utilising organic agriculture as a tool for achieving gender equality in India on the understanding that, unlike traditional agriculture which marginalises women, organic agriculture can empower women. For this purpose the authors interviewed 180 respondents in Uttarakhand, the first state to take up organic agriculture in India. They observed that policy interventions from the state government were encouraging women's participation on a formal basis. They also indicated the need for further studies to compare conventional and organic systems on the basis of gender.

Selvi et al. (2012) addressed the issue of how organic farming has helped in making agriculture environmentally friendly through the use of appropriate technology. Modern agriculture has seen a proliferation of high yielding seeds, chemical fertilizer, irrigation water, pesticides etc. with a view to achieving short-term objectives like provision of food security and generation of foreign exchange. But this has come at the expense of quality of environment and sustainability of resources. In such a situation the authors suggest organic farming as a viable alternative means of agriculture. They argue that organic farming has the potential not only for reducing the cost of cultivation, but also for addressing productivity issues.

By and large, the research activity on organic farming within India has consisted of descriptive work with hardly any exploratory or empirical efforts. Thus, it is obvious that compared to the international activity this research has a long way to go. In particular, as with research in any other topic, there has been a paucity of research with regard to north eastern India in general and Sikkim in particular.

However, in their defence it can be said that organic farming is a much more recent phenomenon within India, and so it can be expected that with time we shall witness much more in-depth efforts in this direction.

3. Objectives of the study

The broad objective of the study is to examine the implications for sustainable livelihood in the context of the organic farming initiative in Sikkim.

Specifically, the study intends to look at the following:

(a) to understand the phenomenon of organic farming in east Sikkim
(b) to determine whether organic farming is acting as a means for sustainable livelihood
(c) to investigate whether people are utilising the facilities provided by the government for organic farming activities
(d) to understand whether the cumulative effect of the above is bringing about a change in the attitudes of the people with regard to earning a livelihood
(e) to understand whether community based organic farming practices in Sikkim are leading to the empowerment of women
(f) to examine the extent of livelihood diversification generated by the pursuit of organic farming and the associated trade-offs
(g) to examine the consumer's willingness to pay for organic food products in view of their environmental as well as substantial health benefits.

4. Database and methodology of the study

The study was carried out in two villages in east Sikkim, Samdur and Sajung, providing two unique perspectives on the phenomenon of organic farming.

At Samdur most of the organic cultivation is done on an individual basis, by and large for family consumption rather than for selling purposes. Owing to their laziness and lack of awareness the cultivators do not utilise the facilities provided by the government. As their expectations of earning immediate returns from their farming activities are not satisfied, they tend to be discouraged from continuing with this kind of activity.

In contrast, Sajung presents the organisation of groups of growers. Cultivators in this area have been organised under the banner of the Krishi Samrudhi Group (KSG), the brainchild of Karma Dwichen Lepcha, a resident of the area. The initiative was started as an effort to involve the women folk in some kind of productive work.

In order to examine the extent of livelihood diversification we have utilised the household livelihood diversification as indicated by the inverse of the Hirchman-Herfindahl (HH) index. This index is then regressed against the variables like age (AGE), educational qualification (EQ), and number of working members (WORKERS) per family. It may be noted that livelihood diversification is positively related with the number of working members in each family as well as education, but has an inverse relationship with age. This is not difficult to understand since the greater the number of working members in a family, the greater the expectation of diversity in occupations. Such diversification will obviously minimise the risk of failure involved in pursuing a single income-based activity. As far as age is concerned, younger people will be more energetic and enthusiastic and can easily change from one job to another. Lastly, the higher the level of education of an individual, the greater the ease with which he or she can succeed in a variety of jobs.

Next, we used the OLS method for studying the effect of variables like educational level, experience and utilisation of government facilities available for organic farming on the per capita income (PCI) of individual growers as well as group-based farmers.

Thirdly, we have considered the customers' willingness to pay (WTP) a premium for organic products in terms of logistic regression where the WTP has been taken in the form of the probability of the customer agreeing to pay a higher price for the organic products. Thus, the WTP can have a minimum value of 0 (in the case of an impossible event) and a maximum value of 1 (in the case of a certain occurrence). This value of the WTP is then regressed against average organic price premiums charged (AP), per capita income (PCI), quality attributes (Q), educational qualifications (EQ) and number of family members (FM). For this purpose forty respondents (buyers) were considered in east Sikkim as part of a market survey where pre-structured questionnaires were utilised to get their views regarding five selected organic food products (vegetables).

Lastly, the contingent valuation method was employed for evaluating the environmental benefits of organically cultivated products utilising the least square method. This method can come into play in the absence of a market that helps to assess the actual value of benefits and costs of changes in environmental quality. It involves directly asking respondents on their willingness to pay (WTP) and willingness to accept (WTA). Structured questionnaires can be utilised for eliciting the relevant information from the respondents.

The dependent variable here is once again the WTP, though in this case it represents the average organic price premiums for the selected food products and not the probability value, as in the previous case. The independent variables consist of PCI, EQ and FM as before, with the addition of age (AGE).

5. Summary of the results and discussion

5.1. Results of analysis for group based organic farming

In order to examine the proposition that participation in group-based farming activities may lead to betterment of the economic condition of the participants, we considered changes in family income and savings with the help of the z test.

(a) Analysis of the change in family income of the respondents

People earn money to satisfy their basic needs. Depending upon their educational qualifications, they settle in a particular occupation and earn income accordingly. For analysing the family income of the respondents, before and after joining the group, the data have been collected and are presented in Table 14.1 below.

Table 14.1. Comparison of the family income of the respondents before and after joining the group

Sl. No.	Monthly Income (Rs.)	Before joining KSG	After joining KSG
1	0–5,000	19	4
2	5,000–10,000	12	11
3	10,000–15,000	1	12
4	15,000–20,000	0	5
5	20,000–25,000	0	0
Total		32	32

Source: Primary Data

Let us take the null hypothesis that there is no significant difference in the average family income of the respondents before and after joining the group, against the alternative hypothesis that the mean family income of the respondents before joining the group is less than that of the mean family income of the respondents after joining the group. Thus, we can say

H_0: $\mu_1 = \mu_2$, against alternative, i.e., H_1: $\mu_1 < \mu_2$

Table 14.2. Particulars of family income before and after joining group

Particulars	Before joining	After joining
Mean (in Rs.)	3,953.125	9,787.5
Standard deviation (in Rs.)	2,616.201	3,941.078
Sample size	32	32

Since the calculated value of $|Z| = 6.978$ is greater than the critical value of 1.64 at a 5% level of significance, the null hypothesis is rejected. Hence, it is inferred that there is a significant difference in the family income. In other words, it also implies that the mean income of the respondents has increased after joining the group.

(b) Analysis of the change in bank savings of the respondents

Savings represent the key factor of each KSG. Savings in the bank are safe and secure; thus, banks offer the best mode of savings. To analyse the saving details of the respondents before and after joining KSG, consider Table 14.3 below.

Table 14.3. Comparison of the change in bank savings of the respondents before and after joining the group

Amount of savings (In Rs.)	Before Joining (No. of Respondents)	After Joining (No. of Respondents)
0–1,000	28	13
1,000–2,000	4	7
2,000–3,000	0	6
3,000–4,000	0	2
4,000–5,000	0	1
5,000–6,000	0	2
6,000–7,000	0	0
7,000–8,000	0	1
	32	32

Source: Primary Data

Let us take the null hypothesis that there is no significant difference in mean bank savings of the respondents before and after joining the group, against the alternative hypothesis that the mean savings of the respondents before joining the group are less than that of the mean savings of the respondents after joining the group. Hence,we can say

H_0: $\mu_3 = \mu_4$, against alternative, i.e., H_1: $\mu_3 < \mu_4$

Table 14.4. Particulars of bank savings before and after joining group

Particulars	Before joining	After joining
Mean (in Rs.)	325	1709.375
Standard deviation (in Rs.)	376.7432599	1656.095169
Sample size	32	32

Since, the calculated value of $|Z| = 4.611$ is greater than the critical value of 1.64 at a 5% level of significance, the null hypothesis is rejected. Hence, it is inferred that there is a significant difference in the savings. In other words, it also implies that the mean savings of the respondents have increased after joining the group.

5.2. Results of analysis for livelihood diversification

The following model is used for OLS regression

$$HHLD_i = a_0 + a_1\,(WORKERS)_i + a_2\,(AGE)_i + a_3\,(EQ)_i + e_i$$

Table 14.5. Livelihood diversification—regression results

Dependent Variable: HHLD				
Method: Least Squares				
Included observations: 30				
Variable	**Coefficient**	**Std. Error**	**t-Statistic**	**Prob.**
WORKERS	0.269778	0.119331	2.260758	0.0324
AGE	-0.002831	0.009384	-0.301720	0.7653
EQ	0.011221	0.033351	0.336462	0.7392
C	1.005685	0.481752	2.087559	0.0468
R-squared	0.181310	**F-statistic**		1.919354
Adjusted R-squared	0.086846	**Prob (F-statistic)**		0.151226

Source: Primary Data

From Table 14.5 it is apparent that all variables except WORKERS are found to be insignificant. The variable WORKERS has the expected sign that its coefficient is positive. This implies that the increase in the number of working members leads to greater livelihood diversification.

5.3. Analysis of the determinants of per capita income from organic farming for both individual growers and group members

We now look at the effect of variables like educational qualification (EQ), experience (EXP) and utilisation of government facilities (GFU) on the per capita income (PCI) of both individual growers and group participants.

The model employed for this purpose is as follows:

$$PCI_i = \alpha_0 + \alpha_1 (EQ)_i + \alpha_2 (EXP)_i + \beta_1 (GFU)_i *d_1 + \beta_2 (GFU)_i *d_2 + u_i$$

Table 14.6. Results of regression of the determinants of per capita income

Dependent Variable: PCI				
Method: Least Squares				
Included observations: 53				
Variable	Coefficient	Std. Error	t-Statistic	Prob.
C	8188.535	6967.467	1.175253	0.2457
EQ	117.0550	558.1610	0.209716	0.8348
EX	-22.43503	246.7798	-0.090911	0.9279
GOVT*D1	2808.782	769.5725	3.649795	0.0006
GOVT*D2	-2132.805	5496.282	-0.388045	0.6997
R-squared	0.430484	F-statistic		9.070539
Adjusted R-squared	0.383025	Prob (F-statistic)		0.000015

Source: Primary Data

From the results in Table 14.6 it is clear that only one variable is highly statistically significant, i.e. utilisation of government facilities for the group members (as indicated by the highly significant dummy variable d_1 corresponding to group participants, in contrast to the insignificant dummy variable d_2 for individual farmers). This can be explained by the fact that the government provides facilities mostly in a free of cost manner and for some at a subsidy to the market. These are utilised by members of the groups; newer members become aware of the availability of such facilities from existing members. On the other hand, individual members lack the motivation to make use of such facilities. The variables like education and experience are highly insignificant; in other words, these factors are observed to have no tangible impact on PCI.

5.4. Analysis of the consumer's willingness-to-pay for organic food products in view of their environmental and health benefits

The evaluation of the extent of the customers' willingness to pay a premium for organic products is to be carried out with the help of a logistic regression. The proposed model for this purpose is as follows:

$$Lij = \ln(P_{ij} / 1 - P_{ij}) = \beta_1 + \beta_2 APjk + \beta_3 PCIi + \beta_4 Qi + \beta_5 EQi + \beta_6 FMi + Ui$$

where

Lij = Whether the ith respondent is willing to pay a price premium for the j selected food product or not
APjk = Organic price premiums (average) charged for any of the jth selected product at the k sampled stores
PCIi = Per capita income of ith respondent
Qi = Quality attributes perceptions of ith respondent
EQi = Educational qualification of the ith respondent
FMi = Number of family members of the ith respondent
Ui = Error term

Table 14.7. Regression results of the determinants of the customers' willingness to pay a premium for organic products

Dependent Variable: Lij				
Method: ML - Binary Logit				
Convergence achieved after 6 iterations				
Variable	**Coefficient**	**Std. Error**	**z-Statistic**	**Prob.**
Avg Price	0.336684	0.152323	2.210323	0.0271
PCI	0.000339	0.000237	1.432218	0.1521
Quality	2.666408	1.205799	2.211321	0.0270
EQ	-0.236413	0.140235	-1.685833	0.0918
FM	-0.190179	0.378380	-0.502613	0.6152
C	-15.11278	7.051145	-2.143309	0.0321
McFadden R-squared	0.440988	LR statistic		19.83865
Prob (LR statistic)	0.001340			

Source: Primary Data

From the Table 14.7 it is clear that three variables have a significant positive impact on WTP: average price, PCI and quality perceptions of the respondents, as suggested by the positive values of the respective coefficients and the corresponding probability values.

The interpretation for this is that in spite of the increase in the average prices people may be willing to pay more for organic products due to the associated health benefits. With the increase in PCI there can be an expected increase in the tendency to utilise such products as the buyers will now have greater capacity to absorb the increase in prices. Again, the greater health benefits expected by the buyers will be reflected through the perceptions of the quality attributes.

On the other hand, education has a significant but negative relationship with WTP, as suggested by the negative coefficient. This is the opposite of what would be expected in reality. In other words, with higher levels of education it is usually assumed that there would be a greater preference for organic products.

The remaining regressor (FM—family members) is found to be insignificant.

However, together all the regressors have a significant impact on the WTP as the LR statistic is 19.83865 with a p value of 0.001340, which is very small.

5.5 Results of contingent valuation method in examining the environmental benefits of organically cultivated products

Apart from the logistic regression, the contingent valuation method has also been used for examining the benefits of organically cultivated products and the proposed model is as follows:

$$WTPi = \beta_1 + \beta_2 PCIi + \beta_3 EQi + \beta_4 FMi + \beta_5 AGEi + Ui$$

where

WTPi=Average organic price premiums for the selected food products
PCIi=Per capita income of ith respondent
EQi=Educational qualification of the ith respondent
FMi=Number of family members of the ith respondent
AGEi=Age of the ith respondent
Ui=Error term

Table 14.8. Regression results for the contingent valuation method

Dependent Variable: WTP				
Method: Least Squares				
Included observations: 40				
Variable	Coefficient	Std. Error	t-Statistic	Prob.
PCI	2.54E-05	0.000199	0.127820	0.8990
EQ	0.304357	0.287790	1.057565	0.2975
FM	1.910316	0.721174	2.648897	0.0120
AGE	-0.321071	0.108851	-2.949629	0.0056
C	50.86989	7.129929	7.134698	0.0000
R-squared	0.358296	F-statistic		4.885566
Adjusted R-squared	0.284958	Prob(F-statistic)		0.003090

Source: Primary Data

Among the regressors, only the number of family members (FM) and age (AGE) are found to be significant, but not with expected signs. However, in the case of age the unexpected results can be explained by the fact that the younger generations are more aware of organic products and the associated benefits. They may be more ready to experiment with such products compared to older members who may have more traditional tastes. Similarly, as organic farming has become an accepted practice in Sikkim and the government is also encouraging it through the provision of various facilities, the cost involved may have ceased to be a matter of concern to the growers. Consequently, family size does not act as a constraining factor.

However, together all the regressors have a significant impact on the WTP which is evident from the values of the F statistic and its p value of 0.003090, which is very small. Again, from the values of R^2 and adjusted R^2 it can be said that the model is quite a good fit.

6. Conclusions and Policy Prescriptions

The foregoing study and analysis lead to a number of conclusions.

Substantial changes have been witnessed in the economic profile of the participants as a result of group-based farming. As the majority of these participants are women and as there have been significant changes in their income sand savings, it may suggest that organic farming can act as a vehicle for female empowerment by making them stronger economically.

In general, apart from women, men can also take up this kind of farming as a gainful economic activity as they also experience the changes revealed through the results of the preceding logistic regression.

The results of the logistic regression indicate that WTP has a positive significant relationship with average price premium charged, per capita income and perception regarding the quality attributes of the products, while a negative relationship with education. Again, the contingent valuation method showed that average organic price premiums are influenced by the number of family members and the age of the respondent. The higher WTP among the buyers can lead to long-term demand for these products in the market, which in turn suggests the possibility of sustainable livelihoods for the male members of the families.

Government facilities are the only factor influencing the PCI of group-based participants. The fact that neither education nor experience are found to have an effect on the rate of income generation further confirms this conclusion.

These conclusions are borne out by the interaction with respondents. The survey of sellers carried out as a part of this study revealed that the selected organic vegetables sold in and around Gangtok are sourced from different places in east Sikkim like Assam Linsey, Namthang, Singtam, Rhenock, Rongly, Namchi, Pakhim, Pakyong and Ranka. The demand for such products has increased and customers are actually prepared to pay higher prices for the same. The supply of organic products are mainly seasonal, and this is more so as these products avoid the chemical fertilisers utilised in traditional agriculture.

The government has to take a number of steps to make organic farming both broad based and long-lasting.

Firstly, the government needs to address the problems faced by cultivators in this area like the non-availability of dependable sources of water. Similarly, the government could examine ways to streamline the delivery mechanism for agricultural inputs to the farmers.

It is also necessary to start a major marketing campaign so as to make people aware of the benefits of organic products over more conventionally derived produce. The lack of knowledge about the substantial health benefits of organic products and the significant price difference against traditional products causes people to ignore the former.

While Sikkim is well-recognised as a pioneer in the field of organic farming, there is no effort to exploit this situation with a view to making the state the leading exporter of organic products. The state government should launch concerted efforts to ensure that organic products from the state reach markets in other cities of the country and even overseas, where such products will not only find a dedicated customer base but will also fetch revenue for the state and foreign exchange for the country. The state government could also consider inviting private public partnership (PPP) endeavours to set up dedicated cold chains for transporting the products like vegetables and flowers from the field directly to the markets, so that the customer not only gets farm fresh products at cost effective prices, but at the same time the producers also get proper value for their produce with minimum difficulty.

Here, it would be worthwhile to note that as much as 75% of the agricultural products available in Sikkim are sourced from other parts of India, these being obviously inorganic in nature. Thus, the question of food security becomes paramount for the state. From this perspective it may be necessary to consider the broad basing of organic farming as an alternative, both for providing a more healthy option, as well as a satisfactory solution to the problem of food insecurity.

References

Chouichom, S., & Yamao, M. "Comparing Opinions and Attitudes of Organic and Non-Organic Farmers towards an Organic Rice Farming System in North-Eastern Thailand." *Journal of Organic Systems* 5 (1) (2010): 25–35.

Greene, Catherine & Kremen, Amy. "U.S. Organic Farming in 2000-2001: Adoption of Certified Systems." U.S. Department of Agriculture, Economic Research Service, Resource Economics Division, Agriculture Information Bulletin No. 780, February 2003.

Mandal, Subhasis, Datta, K. K., Hore, D. K. & Mohanty, S. "Biodiversity and Organic Agriculture: Opportunities and Challenges for the North-East Region of India and a Model for the Principles Involved." *Outlook on Agriculture* 37 (2) (2008): 87–94.

Mollá-Bauzá M. B., Martínez, L. M. C, Poveda, A. M. & Pèrez, M. R. "Determination of the Surplus that Customers are Willing to Pay for Organic Wine." *Spanish Journal of Agricultural Research* 3 (1) (2005): 43–51.

Narayanan, S. "Organic Farming in India: Relevance, Problems and Constraints." Occasional Paper 38 for National Bank for Agriculture and Rural Development Department of Economic Analysis and Research, Mumbai, 2005.

Niggli, U., Schmid, H. & Fliessbach, A. "Organic Farming and Climate Change." Technical Paper prepared for Research Institute of Organic Agriculture (FiBL), Switzerland, on behalf of the International Trade Centre, UNCTAD/WTO, 2007.

Reddy, B. Suresh. "Organic Farming: Status, Issues and Prospects—A Review." *Agricultural Economics Research Review* 23 (2010): 343–358.

Rigby, D. & Caceres, D. "Organic Farming and the Sustainability of Agricultural Systems." *Agricultural Systems* 68 (1) (2001): 21–40.

Rodriguez, E., Lacaze, V. & Lupin, B. "Contingent Valuation of Customers' Willingness of Pay for Organic Food in Argentina." Paper presented at the 12th Congress of the European Association of Agricultural Economists (EAAE) 2008.

Selvi, S., Karthikeyan, R. & Vanitha, U. "Organic Farming: Technology for Environment-Friendly Agriculture." Paper presented at the International Conference on Advances in Engineering, Science and Management (ICAESM), Nagapattinam, Tamil Nadu, India March 30–31, 2012.

Heaton, Shane. *Organic Farming, Food Quality and Human Health: A Review of the Evidence*. Bristol, UK: Soil Association, 2001.

Shiva, Vandana. "Organic Farming: The Real Green Revolution for Removal Of Poverty, Hunger And Ecological Insecurity." Paper presented at the Fifth Annual Conference of the German Council for Sustainable Development, Berlin, September 6, 2005.

Subrahmanyeswari, B. & Chander, M.. "Organic Agriculture: A Way Forward to Achieve Gender Equality in India." *Journal of Organic Systems* 6 (3) (2011): 13–19.

Venkareswarlu, B. "Organic Farming in Rainfed Agriculture: Prospects and Limitations." In *Organic Farming in Rainfed Agriculture: Opportunities and Constraints*, eds. B. Venkareswarlu, S. S. Balloli & Y. S. Ramakrishna. Based on invited papers from faculty of ICAR Winter School on "Organic Farming in Rainfed Agriculture" held at the Centre for Research in Dryland Agriculture, Hyderabad, India, November 1–21, 2007.

CHAPTER FIFTEEN

ECOLOGY VS ECONOMY: QUEST FOR LIVELIHOOD IN A CONFLICT ZONE- ANALYSIS FROM CHHATTRISGARH

PRADIP KUMAR PARIDA*
AND NOTAN BHUSAN KAR

1. Introduction

Though the state of Chhattisgarh was only recently created, it has been in the limelight for the last couple of years, particularly in the aftermath of its creation. The primary objective of creation of small states is to basically speed up the process of development, service delivery mechanism and creation of institutions for a small geographical territory which will lead towards a conducive atmosphere for the people to actively participate in the governance system of the state, but it also has lager implications with economic, political, socio-cultural and linguistic dimensions. If we look into the post-colonial history of South-Asia in general, and post-independent India in particular, the history of state creation has always followed a trajectory from Independence until today.

The crux of the problem is whether it created an atmosphere for the solution of the problems for which the state was created. Has it met the aspirations of the people of the region who sacrificed many things for the creation of the new "state?" This chapter tries to examine these dimensions of Chhattisgarh with specific reference to human security in the present context. The creation of Chhattisgarh and its impact on the governance of the state, in terms of service delivery to the people at grass roots level, is the major theme. The tribals in particular constitute the majority of the population and the region is rich in natural resources, i.e. forests, bauxite, dolomite, iron ore and many other minerals, which are technically under the state control. Do these "subalterns" really gain some benefits in comparison to the times "before" and "after" the creation of the state? Rather, the situation has deteriorated to such an extent that there is a division in a peaceful society in the name of "law and order," where the tribal has no role to play. This has been outsourced from state responsibility to some other organizations, though their legal sanctity is being questioned. The people living there are leading voiceless and hopeless lives, like the situation immediately after a natural disaster like an earthquake or a super cyclone. The fundamental difference between a natural disaster and this type of manmade disaster is very important in this context. At least with a natural disaster, where the people do not have any control over it, there is sympathy from outside and a considerable amount of help in post-disaster rehabilitation. What about these types of manmade disasters, where there is an armed conflict and threat to human security as well as loss of life andlivelihood? The state does not recognize it, and rather legitimizes its action in the name of law and order. Who is to be blamed for this? What is the alternative here? How to come out of a situation which carries out a vicious circle of violence? What about the voice of those innocent citizens? And what about the governance system in this context?

What is so specific about Chhattisgarh in the present context?

The existing socio-political scenario of Chhattisgarh has made it a unique case study for interrogating the nature of state, the rights of citizens, the issues of human security, livelihood and life with dignity and the question of governance. In other words, the constitutional provision of India, which talks about the right to life as a fundamental right, is being squeezed in such a manner that the existence of "citizens" and the "public space" for them isquestioned. Though it is not a natural disaster according to the dictionary meaning, it has repercussions in the social environment, living patterns, food habits, livelihoods, ecology and culture of the tribal people, with specific reference to the Bastar region of the state. The consequences of this disaster are categorically visible today. It has literally created a zone of civil war, where tribals kill other tribals, where a LAKRA is killing another LAKRA or a GOND is killing another GOND, in the name of Salwa Judum or the so-called "Spontaneous Peace Movement." There is a mass migration of people from their natural habitats, homes and villages where they have lived for centuries. They have been put together in camps on the road sides, mud thatched houses have been built, and farming is their occupation no longer. They have sometimes been given food rations, but sometimes this has not been available (according to certain fact-finding team reports). It seems as if this is a zone of civil war. The maintenance of law and order, which is basically under state jurisdiction, has been outsourced to a private militia created to counter another non-state militia. It has been argued by the state authorities that this (so-called Salwa Judum) is a spontaneous movement against the naxalite movement (which is a radical left movement against the state and its power).There is little human security in this context.

2. Brief political history of the state

Chhattisgarh was carved out from the existing Madhya Pradesh on November 1, 2001 with 15 districts. The geographical location, administrative convenience and underdevelopment were considered as the factors triggering the demand for the formation of another state, along with the formation of Jharkhand and Uttaranchal. The population of Chhattisgarh is notable for its high proportion of Scheduled Tribes, while some specific sects are primarily constituted of Scheduled Castes. Of the total population of Chhattisgarh, the tribals constitute at least 32.5%, which is a significantly high percentage. In the last few decades, the demographic profile of tribal-dominated areas has undergone a change. This is a cause for concern as it represents the large-scale intrusion of non-tribals in tribal areas. This changing demographic profile is strongly evident in Bastar, where the proportion of tribals has decreased in the last few decades

3. The Geography, Economy and Culture of Bastar

The identity of Chhattisgarh was created and evolved through a complex process that has largely charted its own course. A combination of cultural, historical, social, economic and political factors have contributed to this process. The wide pluralities of cultures, traditions, histories and customs existing in the region have combined to form a unique mixture that has fed into the development of the Chhattisgarh ethos and identity. However, the key point is that the identity of Chhattisgarh cannot be viewed as separate from the people of Chhattisgarh. It is important to note that the Chhattisgarh identity has been asserted in different forms and has become more pronounced in adverse circumstances, manifesting itself especially as protest against exploitation.

Due to its large tribal population, Chhattisgarh has historically not been a part of the mainstream and has therefore remained underdeveloped. Critical indicators for education and health have remained low. However, as stated above, the region was influenced by mainstream traditional Hindu culture as the overarching organising principle, despite the presence of a large percentage of Scheduled Castes and Tribes. This oppressive, hierarchical social and religious order was not accepted, and from the seventeenth century onwards the social history of Chhattisgarh is marked by the process of questioning and protests in the form of a number of socio-religious reform movements. These movements established an identity for Chhattisgarh.

In the nineteenth century a new system of property rights and revenue collection known as the malgujari settlement was introduced in Chhattisgarh. The new system was implemented with the sole purpose of expropriation and exploitation of marginal farmers, sharecroppers and farm servants by the upper caste Malgujars. Satnam Panth and its followers responded to this exploitative system through various strategies. In several cases the Satnamis deserted villages or continued with the process of Lakhabatta or the periodic redistribution of land, despite the implementation of the new system. Their united challenge to the upper caste Malgujars over the issues of rent and loss of land in the last decade of the nineteenth century was a reflection of the solidarity of Satnamis. This form of protest and response to the new system or property rights and malgujari settlements was widespread among the Satnamis of Chhattisgarh.

In 1908 the forests were first made reserved forests and the contractors given rights to take timber and wood for railway sleepers. This deprived the tribals of one of the main sources of their livelihood. Leasing out of liquor monopolies also aggravated the situation, as the locally made country liquor was declared illegal. The introduction of education and schools was seen by the tribals as an attempt by the state to subvert their culture and therefore became a precipitating factor. Finally, the brutality and exploitation by the police culminated in the Bhumkal rebellion.

Naxalism has a long history at Bastar, with a base here from the early 1970s. Gradually, it has taken over a large area. The state government announced that ten districts are affected by this violent movement, particularly the districts which are tribal dominated and are socio-economically worse-off. The exploitation by traders, land alienation, massive deforestation and displacement of people, due to mega projects, lack of participation of tribals in so called development process, unacceptable governance system, lack of basic amenities of civic life like primary school, hospital, village roads, drinking water facilities and employment generation systems, have triggered frustration among the people. Their livelihood, dependent on forests for centuries, is no longer guaranteed due to the depletion of forest resources. In such a situation the exploitation of local politicians, government officials(particularly the police), forest officials and village patwaris (who hold the land records of the locality) have added fuel to the fire of the problems of tribals. All these have cumulatively increased their dissatisfaction with the state. For this reason the tribals found hope in the Naxals, who have shown sympathy for them for decades and on occasion were successful in demanding the dues for the tribals from the government officials, as well as contractors, middle men and so on.

4. The "Salwa Judum" and "human security" problem

Today, in those parts of Chhattisgarh where the so-called *Salwa Judum* movement is going on, tribals had been the worst sufferers. The Naxilites never hesitate to kill tribals as police informers. Police, SPOs and paramilitary forces also torture and kill them in the name of Naxalite sympathizers. Today, there is a commando group from Nagalandat Bastar for the maintenance of law and order. But for these paramilitary forces it is difficult to distinguish between Naxalite sympathizers and innocent tribals. Ethnic composition, cultural factors and even their dress make it very difficult to distinguish between various groups of tribals. Ultimately, the tribals have been sandwiched between these two forces. Now the situation is that a tribal is killing another tribal, a Lakra is killing another Lakra, a Majhi is killing another Majhi, and a Gond is killing another Gond.

What about the "human security" in this context? What about the health and primary education systems? What about agriculture, animal husbandry and the livelihood patterns? What about their right to lives with dignity, a constitutionally guaranteed provision, highlighted by the honourable Supreme Court in many judgments? The schools were closed down and made into camps for the police, wherein there is no school as such. In this scenario, children are the worst sufferers. The Naxalites are not against the education system as such, but they oppose it as schools are the places that can be easily utilized by the police against them. Hospitals were closed down as no government medical staff were ready to go to such a conflict zone, and only some Christian missionary activists and people from Doctors without Borders were present there.

5. The "fundamental rights" of an individual in a given state structure

What are the rights of a "citizen" in such a situation? As a citizen in a "democratic state" there is little space to ventilate grievances within the state structure itself. The Naxalite movement is not simply a law and order problem, but has certain larger socio-political and economic dimensions involved with it. By the way, Naxalites were ideologically in favour of a militant armed struggle against the state to abolish the state structure and transfer the state power. It has its roots in socio-economic issues, underdevelopment, and bad governance systems. Hence, it is also a socio-economic problem which has culminated into a political problem with symptoms like the law and order problem in the present context. It has to be tackled at a political level, which is totally undermined today. There hardly exists any space for dialogue, discussion or debate with those groups who are against the state in armed struggle.

6. Findings

The fundamental question is why this is happening in such a manner today. Though that part of Chhattisgarh is full of natural resources, the tribals are at the bottom of the layers of access to such resources. Hardly any parameter of the so called "development" has reached them, i.e. primary schools, primary health centres, land holding for subsistence agriculture, village networking roads—all these are lacking. A tribal does not feel as though they are a part of the state system. Their alienation from the state and the larger society is increasing day by day, including land alienation. The massive exploitation by petty contractors, forest mafias, and big industrialists usurping land at a nominal price or free of cost are creating suspicion in the minds of a tribal as to who is reaping the benefits. For example, though the Bastar region has an abundance of mineral resources, such as the Bailadila mines and large forest tracts, along with a railway line, it has hardly been of any tangible benefit to the tribals. No industry has been established there, and the iron ore expropriated from Kirandul has been exported to Japan. At present, whatever industry is being established, the land is taken without their permission even if it comes under PESA, and as far as jobs are concerned tribals are only given the status of unskilled labourers.

There is lack of proper implementation of any development programme like the recently enacted NREGA (according to many fact-finding reports from the field, including the report of Prof. Jean Dreeze).It has not been able to deliver any substantial beneficial impact on the tribals of Bastar, who were dependent on forest resources and rivers for centuries, generation after generation, and are now deprived of all these due to increasing state control as well as privatization of water, timber and forest resources. Hence their grievances and frustration, and Naxalism being given the chance to breed. The tribals have found an avenue to express their hopes as well as frustrations. Hence, there is a clear-cut vacuum in the governance system of the state, which is now occupied by Naxalites.

It is a situation of a complex civil war. The state has outsourced its responsibility of maintaining law and order. A private army, called SPOs, is taking care of it in the name of volunteers who will protect their communities. Ultimately, a tribal is killing another tribal, and police are killing tribals in the name of Naxalite supporters. Naxalites kill tribals as police informers. SPOs are killing their own brothers. The society is divided and it is a very complex situation.

Hence, there is a need to theorize all these issues from a multi-disciplinary approach. We must have a paradigmatic basis to understand the phenomena from the sociological, political and economic viewpoints. Economic deprivation is openly visible in the absence of user and access rights to rich natural resources. The fundamental rights of the people guaranteed by the constitution are also violated (i.e. right to lives with dignity and to livelihood) due to economic exploitation. Even some rights may not be guaranteed by the constitution but enjoyed through custom and social sanction. When these are violated the affected people protest against it, often violently. There needs to be a space for dialogue, debate and discussion in this scenario, and that involves a real democratic process.

7. Conclusion and suggestions

It is not merely political violence but also violence with wider dimensions in all spheres of the given society which has camouflaged that particular state today.

The role of the media is extremely important in this context, as it is the local, national and (to a certain extent) international media that can highlight the real plight of the people who are affected by this type of law and order situation. Though there is a possibility of local media being affected by the pressure tactics of the state government or other vested interests, at least the national media can highlight the realities from the ground level. Though some attempts have been made by certain sections of the national media, they have not reflected the total in totality.

Similarly, the role of academia is extremely important against this backdrop. Though some social scientists from certain research institutes and universities from outside the state are trying their level best to understand this situation and highlight the problems of the affected persons, nothing substantial within the state has been attempted from the academic community. The important reason for this is intimidation by the state power. This is unfortunate, as academia is suppressed from its pursuit of truth.

At the same time, the role of the civil society is also extremely useful in bridging the gap between the divided communities and finding an alternative mechanism to bring about the solution. Though some civil society organizations in Chhattisgarh are trying their level best (particularly many local NGOs and other community based organizations, who are working on health, education, livelihood generation etc.), their efforts are far from adequate for removing a problem of this nature. This is however extremely commendable, as some international voluntary organizations have made their presence felt in helping out the affected persons.

The role of the political parties is conspicuous in this context. As a matter of fact, leaders of major political parties of the state are supporting the state government action. The reason is the fear of losing political legitimacy, the so-called state support and "power" involved with it, and also the threat of growing naxalism. However, there are minor exceptions.

There is a need to change the paradigm of public policy to address these types of challenges. Not only Chhattisgarh but also the other tribal belts of the country, i.e. Nagaland, Bodoland, Gorkhaland, Mizoram, Tripura and Jharkhand are witnessing this problem. Whatever might be the circumstances and causes of the problem, it is ultimately the innocent tribals who are the worst sufferers in Chhattisgarh. Human security is affected to a large extent, and the right to life with dignity is undermined at any cost. As an aftermath, the emergence of a neo-liberal regime in the context of globalization is witnessing these types of phenomena all over the world. The reason for this is that when the market economy determines the state policies, the rights of the citizens, particularly human rights, take a back seat. In this context the most vulnerable sections are the subalterns, i.e. dalits, tribals, minorities and other socially marginalized sections. The present scenario of Chhattisgarh is a classic example of this.

References

Burman, B. K. "Draft National Tribal Policy of 2006: Creating Consternation." *Economic and Political Weekly*, Mumbai, Vol. XLI (34) (2006).

Chaturvedi, Vinayak, (ed.). *Mapping Subaltern Studies and the Postcolonial*. London and New York, 2000.

Chakraborty, Sudip. *Red Sun- Travels in Naxalite Country.* New Delhi: Penguin, 2008.

Concerned Citizens Committee Report on Salwa Judum in Chhatisgarh, 2006.

Cronin, Stephanie, (ed.). *Subalterns and Social Protest: History from Below in the Middle East and North Africa.* Routledge, 2008.

Elwin, Verrier. *Nagaland.* Research Department, Advisor's Secretariat, Shillong, 1961.

IRMA report on PESA and Governance in India, IRMA, Anand, Gujarat, 2010.

Ghurye, G. S. *The Scheduled Tribes.* Mumbai: Popular Book Depot, 1959.

Ludden, David, (ed). *Reading Subaltern Studies. Critical History, Contested Meaning and the Globalization of South Asia.* London, 2001.

Majumdar, D. N. *Affairs of a Tribe.* Lucknow: Universal Publishers, 1953.

Planning Commission, Government of India. "Report on Extremist Affected Areas." 2004.

Rao, V. K. R. V. "Social Change and Tribal Society." *Journal of Social Research* IX (2) (1966).

Sachidanad: *Culture and Change in Tribal Bihar.* Calcutta: Book land Limited, 1964.

Sundar, Nandini. *Subalterns and the Sovereign—An Anthropological History of Bastar*, 1854–1996. New Delhi: OUP, 1998.

Supreme Court, High Court and other judgments related to the above mentioned issues, 2000–2010.

Srivastava, S. K. *The Tharus- A Study of Culture Dynamics.* Agra: Agra University Press, 1958.

Tandon, J. S. *The Report of Raurkela Industrial Complex.* (MSS), 1981.

Vidyarthi, L. P. *Ghaghra- A Village of Chotanagpur.* New Delhi: Manager Publications, 1966.

Vidyarthi, L.P. & Binay Kumar Rai. *The Tribal Culture of India.* New Delhi: Concept Publishing Company, 1976.

CHAPTER SIXTEEN

STRATEGIES FOR SUSTAINABLE LIVELIHOOD ENHANCEMENT THROUGH FOREST RESOURCE MANAGEMENT: A STUDY FROM JHARGRAM, PASCHIM MEDINIPUR, WEST BENGAL, INDIA

BANANI GHOSH AND SWAPNA GHORAI

1. Introduction

Sustainable development implies the use of natural resources such that future generations can attain the same level of well-being as enjoyed by the present. The notion of sustainable livelihood involves maintenance of the maximum level of well-being over the long term. The concept gains added importance when people largely associated with extractive occupations interact with their natural environment with the aim of meeting their basic material needs (Foli et al. 1997). People who live inside forests are heavily dependent on them for their livelihood, primarily on a subsistence basis. People in this category are often indigenous in nature or are people from minority ethnic groups. People who live near forests usually get involved in agriculture outside the forest, and regularly use forest products (timber, fuel wood, bush foods, medicinal plants etc.) partly for their own subsistence purposes and partly for income generation. Forest degradation negatively affects the structure or functions of the stand or site, and thereby lowers the capacity to supply products and/or services (FAO 2001). The loss of key elements of an ecosystem can alter the balance between its components and lead to long-term or permanent change. A well-managed forest resource can provide a number of major supports to the livelihood of the people, such as a sustainable supply of forest produce, scope for productive economic investment and tourism generating local employment and environmental services (Banerjee 2004). Sustainable forest management practices have potential consequences on the socio-economic livelihood of the forest fringe communities. Sustainable livelihood through forest management ensures more income, increased well-being, reduced vulnerability, improved food security and a more sustainable use of natural forest resources to ensure similar benefits and productivity in the future. Appaih (2009) shows, with reference to Ghana, that the management of forest resources holds the prospects for a sustainable ecological integrity, while at the same time satisfying conditions for sustainable livelihood. Sustainable forest management ensures the supply of the products needed for the satisfaction of livelihood objectives on a continuous basis and for the maintenance of the integrity of the forestry ecological system. This chapter deals with sustainable livelihood in the context of the Jhargram block situated in the western part of West Bengal, and focuses on the role of the enhancement of sustainable livelihood security for forest management.

2. Relevance of the study

The forest plays three distinct functions: it acts as a safety net, the support of current consumption, and acts as a pathway for subsistence and sustainability. The forest provides livelihood security for the people living in close proximity. More than one billion people in the world depend on the forest for their livelihood (World Bank 2001).The issue of sustainable livelihood in forest fringe communities has been perceived as a critical approach to meeting the conservation and management of forest resources and the communities that subsist on it. The Jhargram block is endowed with a dense forest resource which mostly consists of the Sal forest. Most of the families are tribal, dominated by the Lodha tribe. The tribal communities mostly depend on traditional occupations for their livelihood, which depend on traditional activities based in the forests.

3. Study area

The Jhargram block, an administrative division in the Jhargram subdivision of the Paschim Medinipur district of West Bengal, has been taken as the study area. Jhargram is located at 22°27′N 86°59′E and has an area of 539.64 sq km. The gram panchayats are Bundhgora, Chandri, Chubka, Dudhkundi, Lodhasuli, Manikpara, Nodabahar, Patasimul, Radhanagar, Salboni, Sapdhara, Guibani and Sardiha, with the headquarters at Jhargram. The block is bounded by the Binpur block in the north, Jamboni in the west, Sankrail and Gopiballavpur in the south, and

Medinipur and Kharagpur in the east. The general appearance of the area is gentle undulations with ridges covered by a thick growth of dwarf trees and other scrub. As per the 2001 census, the Jhargram block had a total population of 153,381, out of which 78,362 were male and 75,019 were female. This area is situated in Junglemahal. This is one of the backward areas in West Bengal and India. The financial status of most of the people falls in the middle and lower class categories.

Fig. 16.1. Location of the study area

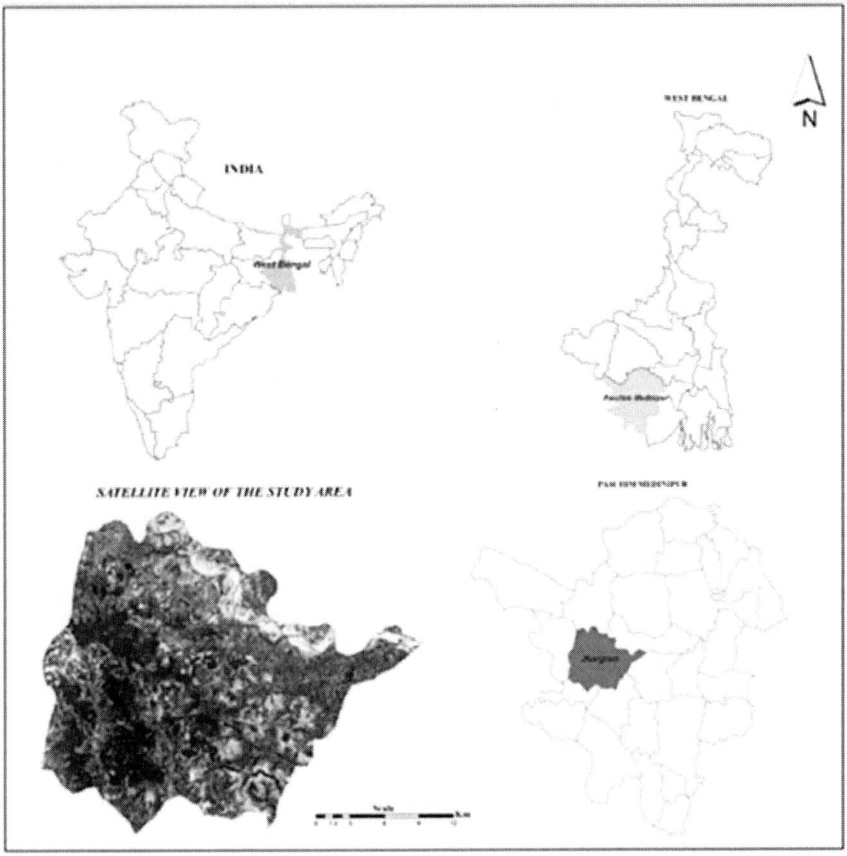

Fig. 16.2. Location of the Jhargram Block

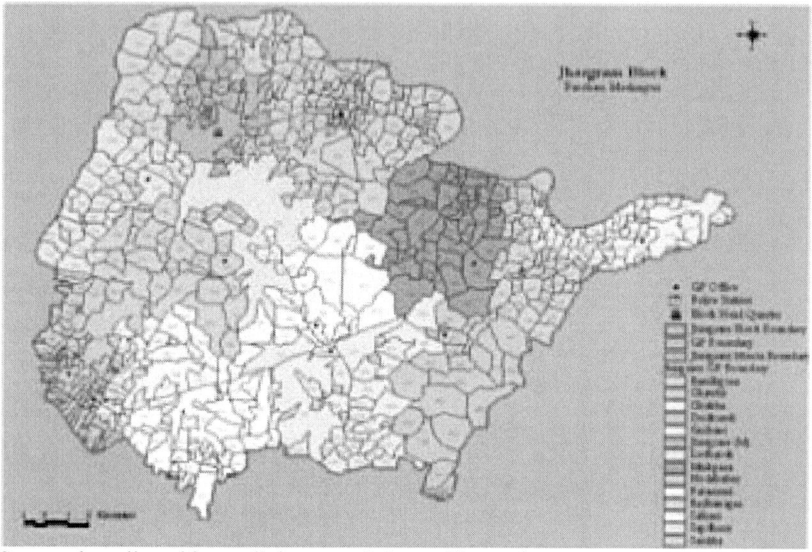

Source: http://paschimmedinipur.gov.in/maps/Show.php

4. Objectives of the Study

- To assess the forest resource base for livelihood.
- To determine the livelihood pattern of the forest fringe villagers.
- To harness the strategy for sustainable livelihood enhancement for forest resource management.

5. Database and methodology

In order to fulfil the above mentioned objectives, some methods have been followed in different stages. These are:

5.1 Pre-field study

- Collection of data from Forest Dept., Census and other various Govt. and Non-Govt. organisations to address the necessary information about forest resources, livelihood and demographic characteristics.

- Collection of information and literature from related books, journals, and websites.
- Collection of maps and satellite images to know the facts of forest degradation and land uses.

5.2 Field study

- A household survey over the study area has been conducted for all sorts of information like socio-economic characteristics, forest produce collection etc. Specific sampling techniques (10% random sampling) have been used to select the households.
- Some qualitative information has also been collected through surveys to get an idea of the quality of life in the study area.

5.3 Post-field

- Different cartographic techniques and GIS platforms are used to feed the results of different analysis.

6. Results and Discussion

6.1 Forest resources and livelihood

The forest (nearly 27% of the study area) assumes a significant role in the material, economic, spiritual and overall livelihood security of the study area. The forest of this block falls a little above the lower limit of sub-group 5B as described in forest types of India by Champion & Seth(1962).Soil types are laterite and red soil. Mainly Sal Jungle is found here, like an island in the red soil-dominated forest area of Jhargram block. The principal species of this area is Sal (Shorea robusta), associated with Peasal (Ptowearpus marsupium), Asan (Terminalia tomenfosa), Rahara (Soymida febrifuga), Dhaw (Anogeissus latifolia), Bahera (Terminalia baferica), Mahul (Madhuca Indica), Vela (Semecarpus Anacardium) etc. Large-size animals are somewhat absent here. The area is one of the backward regions of West Bengal. Forest fringe villagers mainly earn their livelihood from forest resources. A sustainable source of supply of forest produce includes fodder for cattle with downstream economic benefits, firewood for homes and cash sales to outsiders, and poles and timber for agricultural implements, transport carts and house construction (Banerjee 2004).

Minor forest produce plays an important role in the economy of fringe forest villagers, and its collection and marketing is a major source of livelihood. NTFPs can be classified in a number of broad categories according to end use: edible products, fodder for domestic animals, medicines, perfumes and cosmetics, colourants, ornamental articles, utensils, handicrafts, construction materials, and exudates like gums, resins and latex (Shvidenko et al. 2005). The NTFPs are collected seasonally in Jhargram. Mainly firewood is collected for sustaining livelihood throughout all seasons, constituting the major source of energy in rural areas. The ability of the forest to supply biomass energy to the forest fringe villagers is an important economic phenomenon. Its collection is mainly performed by women and they sometimes have to cover long distances for its collection. Branches and twigs are also collected for energy sources when firewood is not available in sufficient quantity. Other NTFPs are Sal leaf, Sal seeds, Kendu leaf, Tasar silk worm, mushroom, fruits and medicinal plants. Sal leaf, Sal seeds and babui grass are marketed after processing by women and children. Babui grass is used for making rope and other necessary products. Different types of fruits continue to be collected through the year. More than 95% of Tasar host plants are confined to forest areas and the tribals are engaged in this to earn some income. A variety of medicinal plants are collected from the forest which include Kalmegh, Arjun (bark), Vela (bark), Bahera (fruits) Satmul (roots), and Iswarmul (roots). Apart from this, vegetables are also collected to supplement the livelihood of the poor households. Traditionally, the forest is a major source of fodder for supporting livestock.

Fig. 16.3. Forest area of the Jhargram Block

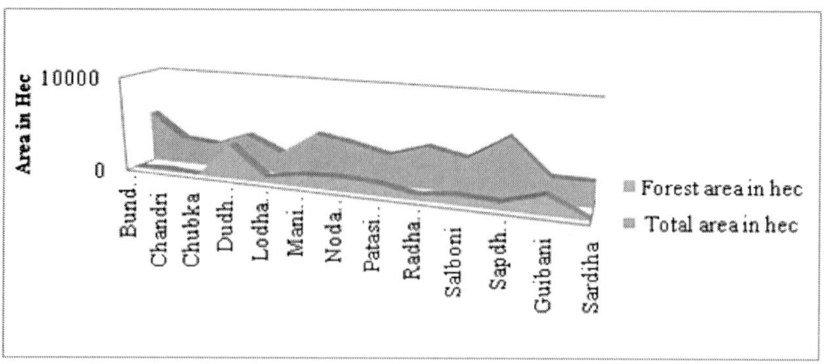

Source: Based on data from Census 2001

Fig. 16.4. Forest cover map of the Jhargram Block 2006

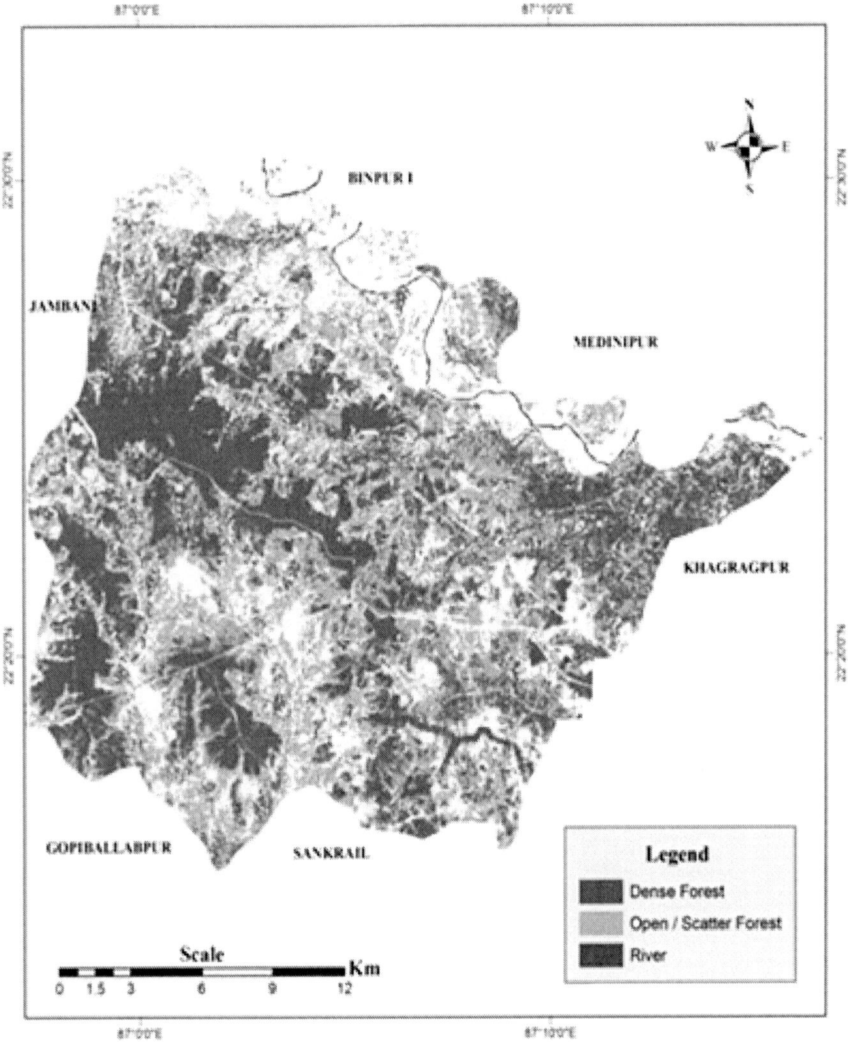

Table 16.1. Different types of NTFPs available in Jhargram Block

Sl No.	Parts of plant collected as NTFPs	Plants
1	Branches and leaves	Firewood
2	Leaf	Kendu
3		Sal
4		Kalmegh
5	Flower	Mahua
6	Seed	Sal
7	Fruits	Bahera
8		Amla
9		Bel
10		Kendu
11	Gum	Sal
12	Bark	Arjun
13		Vela
14	Roots	Iswarmul
15		Satmul

Source: Household Survey 2011–12

6.2 Forest dependence

Forest dependence is mostly related to social class. Landless fringe forest people are more dependent on forest products for their livelihoods. This encompasses (a) people who live within forests and survive on a subsistence basis, often comprising indigenous people or people from minority ethnic groups, and (b) people who live near forests who are usually involved in agriculture outside the forest, and who regularly use forest products (timber, fuel wood, bush foods, medicinal plants etc.) partly for their own subsistence purposes and partly for income generation.

Table 16.2. Collection of some non-timber forest products in Jhargram Block

Month NTFPs	Jan	Feb	Mar	Apr	May	June	July	Aug	Sep	Oct	Nov	Dec
Sal leaf												
Kendu leaf												
Sal seed												
Tasarsilk worm												
Mashroom												
Mahua flower												
Fruits												

Source: Household Survey 2011–12

Fig. 16.5. Relation between the collection of value-added products and distance from forest

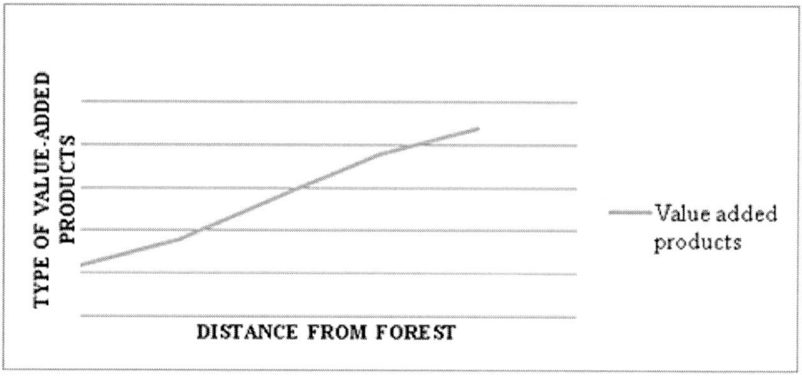

From the household survey, it is observed that 61% of average income of forest fringe villagers is associated with forest resources and 24% from livestock whose production and growth also indirectly depend on the forest in the form of supply of fodder.

Fig. 16.6. Average income of forest fringe villagers from different activities

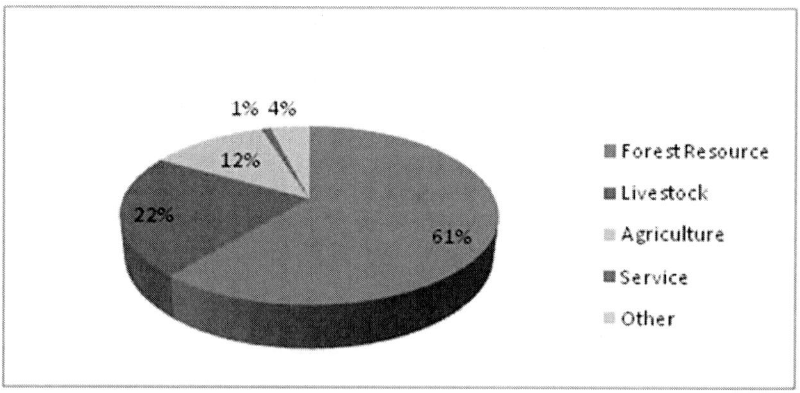

Source: Household Survey 2011–12

Income from NTFPs increased after the introduction of the joint forest management programme. The collection of minor forest produce is done mainly by women and children. There exists an organic relation between forest and fuel. The relatively lower category of people near the forest is associated with the forest more through this relationship. The demand for more valuable forest products is generated by people located in areas away from the forest where other sources of fuel are also available. Processing and marketing of Sal leaf plate is a major economic phenomenon here. On average, 47% of income is generated through making Sal leaf plates.

Fig. 16.7. Average income from different forest resources

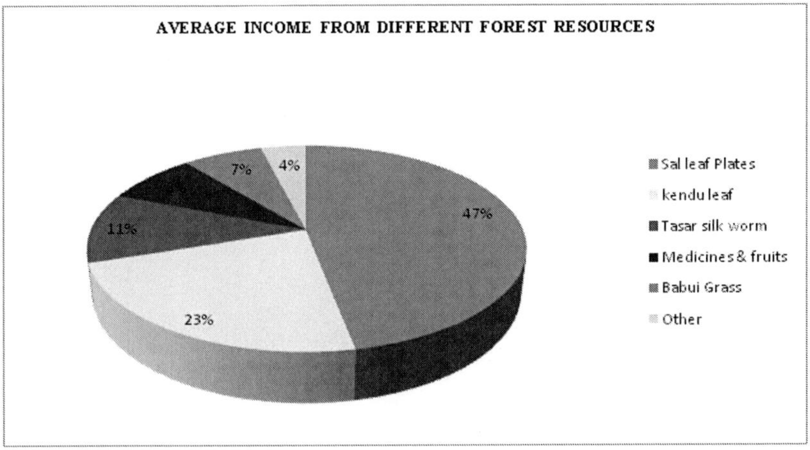

Source: Household Survey 2011–12

Another important use of forest resources is the supply of raw materials for the making of products based on cottage industries such as:

- Kendu leaf—Biri making
- Tasar silk worm—Tasar silk
- Babui grass—rope, mat
- Sal seed—oil, butter
- Mahuya flower—oil, soap, skin care products, syrap
- Wood—Different types of handicrafts

6.3 Forest exploitation and degradation

Degradation of forest has increased in recent years due to heavy biotic interference. A forest cover change map (Fig. 16.8 below) has been produced to show the changes in forests that occurred between 1990 and 2006. Major areas have been identified where large scale changes have occurred. The entire forest of this division is of coppice origin and subjected to maltreatment like repeated felling on short rotation, intensive grazing, lopping etc. The conversion of forest to agricultural land has increased the feed growing number of people. Commercial logging destroys trees similar to the extension of agriculture in forest areas. The cutting of trees for firewood, heavy lopping of foliage for fodder and wanton trampling of saplings by the hooves of domestic animals involved in random grazing have intensified the problems.

Fig. 16.8. Forest cover change map of the Jhargram Block 1990–2006

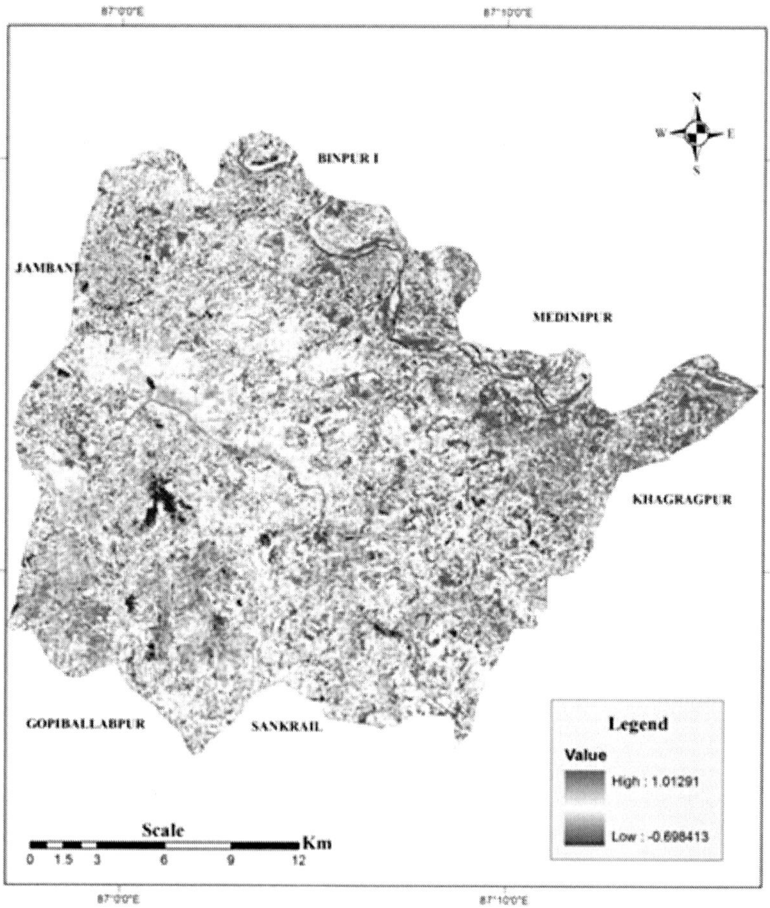

Some indigenous ways of life are threatened by the loss of forests. Tribal women have adjusted to live a traditional lifestyle in the local environment and follow occupations based on natural resources. Therefore, it is mostly the tribal women who face problems in attaining sustainable livelihoods and decent lives due to environmental degradation (Awais et al. 2009). Where forest is replanted, their replacement can mean a loss of quality. The stress of environmental changes may make some species more susceptible to the predation by insects and vulnerable to pollution, disease and fire. Loss of top soil also hampers the productivity of agricultural areas outside forest areas.

6.4 Problems Identified

- Livelihood activity based on NTFPs is highly seasonal in nature.
- Ecological degradation and relatively low accessibility to the forest are major problems for local people.
- Lack of capital and technology for production of value added products near the surrounding area.
- Lack of accessibility to market and organised market facilities.

7. Socio-economic Sustainability and Forest Management: Linking Biodiversity Conservation and Sustainable Livelihoods

Sustainable livelihood is a systemic and adaptive approach that links issues of poverty reduction, sustainability and empowerment processes (e.g. participation, gender empowerment and good governance). A livelihood is sustainable when it can cope with, recover from stresses and shocks and maintain or enhance its capabilities and assets, both now and in the future, while not undermining the natural resource base (Chambers & Conway 1992). Adaptive strategies for sustainable livelihood encourage sustainable use of natural resources and strengthen the society, local institutions and networks so as to create an enabling environment for sustainable livelihood patterns. A livelihood strategy encompasses not only activities that generate income but many other kinds of elements, including cultural and social choices (Ellis 2000).

Degradation of forest leads to shortage of fuel, fodder, foliage and forest produce on which local economies are mostly dependent. A holistic approach for the valuation of forests is essential while examining the issue of compensation for expansion and maintenance of forest cover (Mathur & Sachdeva 2003). Programmes for forest-based communities should be developed depending on poverty, seasonality and other challenges. The importance of resource rights and livelihood security has been emphasized in this context. Natural resource-dependent individuals, households and communities become marginalized because they either have no rights to the resources on which they are dependent or no feasible way to exercise the rights they do have. Promoting and raising awareness of local rights, including traditional rights, are essential, particularly in the face of encroachment from external commercial forces. Formation and stabilization of tribal co-operatives should be effective for this. Programme activities should be launched targeting poor and vulnerable people to ensure that such activities reach poorer groups (including women and girls) and produce tangible livelihood benefits.

In recent years it has become increasingly well recognised that people living in and near forests often have quite sophisticated knowledge and techniques for forest use and management for regulated access to and use of forests. These systems can be described as indigenous forest management systems. An integration of traditional knowledge with modern scientific development is necessary for successful forest management practices. Special emphasis should be given on NTFP production and commercialization in an ecofriendly manner for forest-based economic development. The proper evaluation of NTFPs is important for this. Importance should be given to the production and marketing of different types of handicrafts like mats, carpets, wall hangings and decorative items from the leaf stems and fibres of natural plants. The plans and programmes should be directed towards the upgrading of existing technology support and assistance to infrastructure development and marketing support etc. for betterment of their earning. Efforts should be given to developing the skills of primary collectors and special attention in negotiating with traders and using technology. The livelihood approach may be used to systematically examine differences in the livelihood system across groups that are traditionally disadvantaged and marginalized. This influences their ability to respond to new technologies and market opportunities. Practical knowledge and training in modern techniques should be provided regularly so that their work becomes easy. Women play an important role in the sustainable livelihood development. Considering the involvement of women with the collection of forest resources, and the valuable contribution of women in the management process should be emphasized. A strategic improvement and enhanced provision for their empowerment through capacity building in the context of ecological development are needed.

8. Suggestions and conclusions

Sustainable forest development means the use and conservation of the forest while at the same time improving the economic condition of forest-dependent people. Efforts should be made to motivate people to protect the green treasure as well as ensure their economic betterment. Joint forest management was adopted as an alternative form of management practice. The forest wealth after implementation of the JFM Programme has created a situation for much larger livelihood support in forest-fringe villages. All types of forestry works are being executed by the concerned FPC members. Non-timber forest produces are collected by FPC members and they earn a good amount per family per year. Also, the NREGA programme can be used as a strong weapon for the sustainable development of the people as well as nature if it is used in a judicious, transparent and scientific manner. The loss of forest cover can be compensated through sustainable forest management including protection, restoration, afforestation and reforestation. Increased efforts should be given to preventing forest degradation and enhancing the flow of forest-based economic, social and environmental benefits, including improving the livelihoods of forest-dependent people. Improved co-operation between national, provincial and district counterparts, villagers and relevant NGOs is essential for poverty reduction and promoting sustainable livelihood. The micro-level integrated resource planning, implementation and monitoring through state governmentefforts mayprove extremely useful in this direction.

References

Annual Report of Forest. Jhargram division, Jhargram, Paschim Medinipur 2010–11

Appaih, D. O. "Personifying Sustainable Rural Livelihoods in Forest Fringe Communities in Ghana: A Historic Rhetoric?" *Journal of Food, Agriculture, & Environment* 7 (3–4) (2009): 873–877.

Awais, M., Alam, T. & Asif, M. "Socio-Economic Empowerment of Tribal Women: An Indian Perspective." *International Journal of Rural Studies* 16 (1) (2009).

Banerjee, A. K. *Participatory Forest Management in West Bengal: A Review of Policies and Implementation.* Understanding Livelihood Impacts of Participatory Forest Management Implementation in India and Nepal.Working Paper 3. 2004

Chambers, R. & Conway, G. "Sustainable Rural Livelihoods: Practical Concepts for the 21st Century." IDS discussion paper 296. Brington, IDS 1992.

Ellis, F. "Rural Livelihoods and Diversity in Developing Countries." Oxford: Oxford University Press, 2000.

FAO. "Global Ecological Zoning for the Global Forest Resources Assessment 2000." Forestry.Working paper 56, Rome, 2001.

Foli, E. G., Adate, K. and Agyeman, V. K. (eds.). "Pilotive Collaborative Forest Management Systems for Off-Reserve Areas in Southern Ghana." ITTO/FD Seminar on Sustainable Timber Production from Outside Forest Researves, 73. 1997

Hegde, N. G. "Management of Natural Resources for Sustainable Livelihood—BAIF's Approach In Natural Resources Management and Livelihood Security: Survival Strategies and Sustainable Policies." Eds. K.V. Sundaram, M. Moni and M. M. Jha, Bhoovigyan Vikas Foundation, New Delhi, 1–17. 2004.

Mathur, A. S. & Sachdeva, A.S. *Towards an Economic Approach to Sustainable Forest Development. Perspective Planning Division* Planning Commission, Govt. of India. Working Paper Series. Paper No. 2, 2003.

Panigrahi, N. 2005. Development of Ecotourism in Tribal Regions of Orissa: Potentials and Recommendations. *CEWCES Research Paper*, Paper 9. http://epublications.bond.edu.au/cewces_papers/9

Shvidenko, A., Barber, C. V. and Persson, R. 2005. *Forest and Woodland Systems* (Current state and Trends Assessment). Millennium Ecosystem Assessment, World Resource Institute: Washington, USA.

World Bank. *A Revised Forest Strategy for the World Bank Group.* Washington DC, 2001.

CHAPTER SEVENTEEN

NON-FARM LIVELIHOOD DIVERSIFICATION AND RURAL DEVELOPMENT: EVIDENCE FROM A FIELD SURVEY IN WEST BENGAL

SUCHISMITA MONDAL SARKAR

1. Introduction

A review of a complex archeology of ideas and practices reveals that livelihood diversification has been a survival strategy of rural households in the developing economies to stabilize their income. However, the ability to engage in diverse activities is often governed by their relative ability to access productive resources apart from the socio- economic as well as geographical environment. Even with modernized agriculture there is surplus labour in the developing economies. In most of them the rural sector has been subjected to diminishing returns. Agriculture has been too much over-burdened with the pressure of activities. Moreover, it is increasingly becoming clear that the agricultural sector alone cannot be relied upon as the core activity for rural households as a means of improving livelihood and reducing poverty. One phenomenon gaining prominence in the rural development literature is the promotion and support for non-farm diversification opportunities, particularly in its role in generating productive employment, contributing to economic growth, improving income distribution, and alleviating poverty in rural areas.

Multiple motives force people to diversify their assets, incomes and activities. However, income diversification is not synonymous with livelihood diversification. Income diversification is one of the components of livelihood strategies (Ellis 1998). Livelihood diversification encompasses social institutions, rights, relations and other non income support systems that sustain a living. It consists of a proper blending of farm and non-farm income at the household level that provides resilience against adverse situations.

Diversification of the sectoral activities is conducive in bringing about the desired transformation of the economy in general and rural economy in particular. Transformation means movement of the economy from one stage to another stage of development. As an economy advances technologically over time, the importance of the farm sector in terms of its share in GDP and share in total employment is reduced, and the share of the other two sectors increase gradually. In the rural economy agriculture loses its prime importance over time and livelihood related activities shift in favour of the non-farm sector through transforming, following the path: Farming / Agriculture → Manufacturing / Industry → Services. The role of the rural non-farm sector becomes important as it can provide diverse employment opportunities to the rural people and in the process transform the rural economy in the desired direction of inclusive growth.

2. Diversification and rural development: a conceptual framework

Economic development is often equated with the structural transformation of the economy where the relative share of the agricultural sector in both the national income and labour force declines while that of the industrial and service sector increases. That is, the economic diversification of occupational structure and the relative changes in the employment and income form an integral part of the socio-economic process underlying such transformation. However, the concept of rural development has changed its paradigm. In integrating policies with rural development, the elements of conventional paradigms are often contrasted with alternative paradigms

Table 17.1. Conventional vs. Alternative Factors in Rural Development Policy

Conventional paradigms	Alternative paradigms
Economies of scale	Economies of scope
Specialisation	Diversification
Productivity of labour	Value Added

The pace and the intensity of change in the rural areas signal the reframing of the whole system. This alternative development paradigm is based on needs, local self-reliance, ecology and structural transformation to enable self-management. The issues concerning rural development are largely centred on iniquitous income, opportunities and access of its populace. The explanation for diversification to non-farm activities is the existence of economies of scope in production. Economies of scope exist when the same inputs generate greater per-unit profits when spread across multiple outputs than when dedicated to any one output. The concept differs from that of economies of scale, in which per unit profits increase as the amount of all inputs to production grows. Economies of scale tend to favour specialization. Based on these alternative paradigms of development, we conceptualize our framework as—Economies of scope promote diversification to different non-farm activities, which adds to the income of the household, empowering the rural population. It leads to the decreasing dependence on agriculture and shifting to the non-farm sector which finally paves the path towards transformation of the rural economy, eventually contributing to development. Thus, in conceptualizing the transformation we follow the framework below:

Fig. 17.1. Causation among diversification, empowerment and development: a conceptual framework

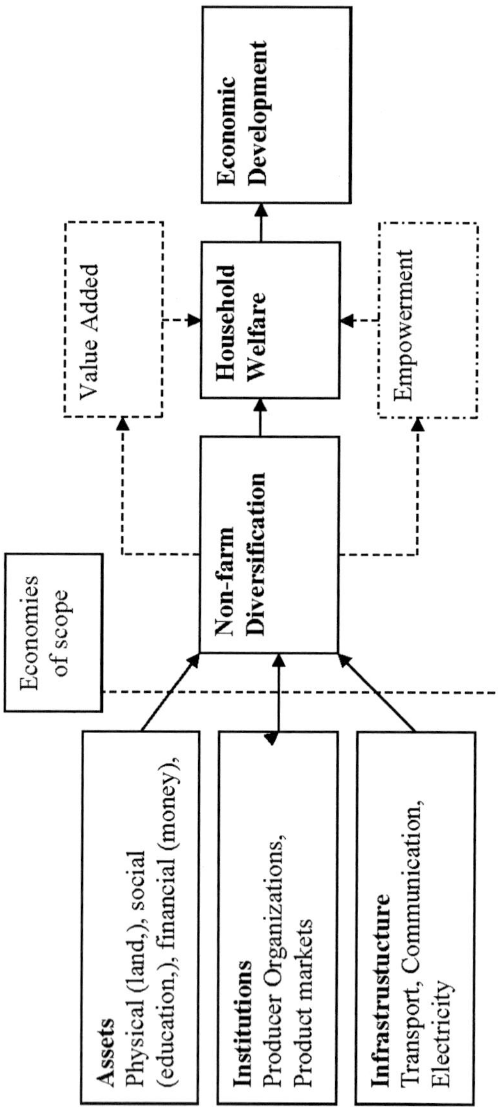

The basic framework is that of a household's portfolio of activities and their impact on welfare. The existence of economies of scope is based on certain incentives and constraints created through the resources available to a household embodied in the assets, institutions and infrastructure. By assets, we are referring to the physical, social, financial and human resources available to the household. Changes in the portfolio of assets, their productivity and the extent to which households have access to them are the attributes that are critical in determining livelihood diversification and ultimately household welfare. The limitations in access to credit and lack of education act as constraints to development. The presence of well-functioning markets helps in reducing transaction costs and risks being involved in non-saleable outputs. Again, access to rural infrastructure including the presence of local markets, usable roads, electricity, telecommunications, etc. provides returns of economic investments. This creates direct and indirect non-farm employment and income opportunities for the poor to improve their welfare. The idea of institutions could be defined as to include many components particularly important for rural farm households, namely producer cooperatives, markets, extension services and the sharecropping system. As the household welfare is enhanced it contributes to the overall development of the region, thereby empowering the rural population.

Diversification takes place due to the simultaneous operation of both distress and developmental factors. At the household level, access to land, household size, caste, gender and educational status are some of the factors that have a bearing on the nature and extent to which people divert. For a rural household, diversification is a means of enhancing household income, thereby overcoming vulnerability. Thus, rural livelihood diversification is considered a derivative of a broader social, economic and political process. The process of diversification away from (outside) agriculture helps the transformation of the rural economy. It: (i) increases the income of the rural people considerably as a non-farm wage is usually higher than an agricultural wage; (ii) provides security and reduces the risk and uncertainty associated with farm income; (iii) reduces the pressure of labour on land; and (iv) reduces the tendency of the rural people to migrate to urban areas. Thus, diversification has become the key to development through the transformation of the rural economy. To analyse the benefits of diversification, proper investigation about the nature, determinants and constraints of diversification is needed, on the basis of which proper policies can be framed to facilitate the process of rural transformation.

Against this backdrop, this study tries to enumerate the determining factors that influence diversification to non-farm jobs and also tries to find out the factors that act as constraints to diversification to the non-farm sphere. The broad objectives of the study are:

(1) To provide a view of the socio-economic aspects of the households undertaking non-farm activities.
(2) To examine the potential factors contributing to rural non-farm diversification and assess the constraints to its growth considering various socio-economic, technological, infrastructural, institutional and policy factors both at the household and regional levels.
(3) To analyse the households' changing perceptions towards their living standards as an impact of diversification into non-farm activities that contribute to the welfare and development of the region.
(4) To identify potential sources and suggest appropriate strategies and policies for the government and non-government rural development organizations in improving and strengthening the capabilities of the rural people economically dependent on the non-farm sector.

3. Data and methodology

Primary data were collected and analysed to arrive at results and conclusions. For the purpose of primary data, a household survey was conducted in the Burdwan II block of the Bardhaman district of West Bengal. The block Burdwan II has an area of 179.14 km^2. It consists of nine gram panchayats: Baikunthapur–I, Barasul–I, Kurmun–II, Baikunthapur–II, Barasul–II, Nabastha–I, Bandul–II, Gobindapur and Nabastha–II. Of these nine gram panchayats only two, Baikunthapur–I and Barasul–II, were selected for the present study. Again, four villages, Tant Khanda and Baje Salepur from Barasul–II and Jotram and Kandorsona from Baikunthapur, were surveyed. Fifty percent of the households from each village were selected on a random basis. The relevant data were collected on the basis of a structured schedule. The interview questionnaire was designed keeping in mind the objective of our survey, covering all the relevant information required for the purpose.

The analytical techniques employed in the study include inverse of Hirschman-Herfindahl Index (HHI) as a measure of diversification, OLS Multiple Regression to identify the major factors influencing non-farm diversification, and the Binary Logistic Regression model to capture the impact of non-farm diversification on the perceived standard of living of our sample households. The diversification index is measured by taking the inverse of the Hirschman Herfindahl Index (HHI). HHI measures the degree of concentration. Therefore, the inverse of this index gives us the measure of diversification. The higher the value of the diversification index implies a greater tendency of the people to diversify. The formula is given as follows:

Diversification Index (DI) = 1 / HHI

If a household is fully concentrated in only one activity, then the index becomes 1. As the value rises, it indicates a higher degree of diversification.

4. General profile of the surveyed households

The socio-economic characteristics of a household are important factors influencing the decision to diversify towards non-farm activities by affecting their ability and motivation to diversify. The four villages studied are situated within 10 km from the main town of Burdwan and are well-equipped with road connectivity and communication facilities. In Table 17.2 below, the combined data show that 50% of the population belongs to the SC category. People belonging to the ST and OBC categories are negligible here. General and Minority population shares almost equal proportions (24.55% and 20.91% respectively).The area is Hindu dominated (about 80% of total population). The average family size is between 5 and 8 in all villages. Almost 70% of the population comprises people aged between 15–60. That means the largest proportion of population belongs to the working class. Data show that about 15% of our population is illiterate, about 15% have reached primary school, while 60% have reached high school and about 10% are graduates or more educated. Therefore we can say that more than 70% (60.08% + 10.08%) of the population has reached the high school education or above. This gives us quite a bright educational scenario though there still exists 15.23% illiteracy in the area. Most of the households have pucca houses and pucca toilets. The tube well is the main source of drinking water and wood is the main source of fuel in the region. Most of the villagers are either landless or marginal farmers. However, the quality of land is quite good and most of the land is well-irrigated. Potato is the highest agricultural revenue-earning crop, with rice and mustard acquiring the second and third places, respectively. But, in spite of having so fertile and well-irrigated agricultural land, due to the falling profitability of agriculture people are switching to different non-farm activities like non-agicultural labour, business, service and others for their alternative livelihood options.

Table 17.2. Socio-economic characteristics of the study area

Background Characteristics	Percentage	Number	Background Characteristics	Percentage	Number
Sex			**Toilet type**		
Male	52.65	268	Pucca	72.73	80
Female	47.35	241	Semi-pucca	12.73	14
			Kachcha	10.00	11
			No facility	4.55	5
Religion			**Sources of drinking water**		
Hindu	79.09	87	Municipal tap	20.91	23
Muslim	20.91	23	Tube well	75.45	83
			Well	3.64	4
Caste			**Sources of fuel**		
General	24.55	27	Wood	65.45	72
SC	50.00	55	(collected or purchased)	0.00	0
ST	1.82	2	Kerosene	20.91	23
OBC	2.73	3	Coal	13.64	15
Minority	20.91	23	LPG		
Age			**Source of income**		
0-6	7.27	37	Cultivation	18.28	Rs. 295882
7-14	11.00	56	Agrl. labour	1.28	Rs. 20683
15-60	72.89	371	Non-Agrl. labour	6.99	Rs. 113083
60 above	8.84	45	Business	47.62	Rs. 770672
			Service	18.96	Rs. 306800
			Others	6.88	Rs. 111322

Family size			**Operational land holding**		
1-4	46.36	51	Landless farmer	32.73	36
5-8	50.00	55	Marginal farmer	42.73	47
9-12	3.64	4	Small farmer	15.45	17
			Medium farmer	3.64	4
			Large farmer	5.45	6
Educational level			**Main agricultural products**		
Illiterate	15.23	74	**(market value in)**		
Primary	14.61	71	Paddy		
High School	60.08	292	Potato	32.05	Rs. 2110725
Graduate and above	10.08	49	Mustard	63.86	Rs. 4205250
				4.08	Rs. 268800
House type			**Asset value**		
Pucca	65.45	72	Livestock	1.73	Rs. 801810
Semi-Pucca	24.55	27	Agrl. asset	90.63	Rs. 42108200
Kachcha	10.00	11	Non-agrl. asset	7.65	Rs. 3552300

Source: Primary Survey 2011

Petty trades like making rice from paddy, door-to-door garment selling, small grocery shops and fast food stalls are the major areas providing non-farm income shares of the households. Migration is a very rare phenomenon in the region. The total assets of a household comprise livestock, agricultural assets and non-agricultural assets, with agricultural assets comprising the highest share. The majority of our sample households are found to have taken loans from formal sources, and most of the loans are taken for business. For agricultural purposes, people are equally dependent on both formal and informal sources of credit. Among the different government policies for rural development, the most widespread in the area is the 100 Days Work Programme under the Mahatma Gandhi National Rural Employment Guarantee Scheme. Among the kinds of activities under the programme, digging ponds and making drains are mostly done to serve the basic irrigation of the villages. Besides these, making roads and plantations are also done occasionally. But employment provided under the scheme is not at all sufficient. People had only 11 days of employment on average in the last year, and the average earnings received by a household were Rs. 1,282. Thus, the effectiveness of the policy has become questionable in this particular region.

5. Multiplicity of household livelihood strategies

We now analyse the patterns of household livelihood strategies. Table 17.3 below shows the distribution of households according to the number of livelihood strategies they are engaged in. It becomes evident that diversification varying in degree has been an important instrument of earning livelihood. Households typically diversify their portfolios of earnings and tend to participate in a number of different activities simultaneously or sequentially. A multiplicity of sources of livelihoods not only allows the households to diversify their sources of earnings; it also helps them to effectively utilize some of their slack resources, such as labour of household members in off-peak seasons and beyond normal working hours.

In order to focus on the issue of diversification we have categorized the households into three broad categories: (i) households that are completely dependent on farming, (ii) households that are engaged in both farming and non-farm activities, and (iii) households that have only non-farm occupations.

According to Table 17.3 below, pure farm dependence is lagging far behind compared to the other categories in all villages. There is a substantial dependence of the households on non-farm activities, as these

Table 17.3. Multiplicity of livelihood in selected villages

	Village Name				Combined Villages
	Tant Khanda	Baje Salepur	Jotram	Kandorsona	
No. of households with only farming	3 (11.53%)	7 (16.28%)	5 (29.41%)	6 (25.00%)	27 (24.55%)
No. of households with farming and one more occupation	1 (3.85%)	10 (23.26%)	5 (29.41%)	8 (33.33%)	24 (21.82%)
No. of households with farming and two more occupation	0 (0.00%)	10 (23.26%)	2 (11.76%)	5 (20.83%)	15 (13.63%)
No. of households with farming and three more occupation	0 (0.00%)	2 (4.65%)	0 (0.00%)	4 (16.67%)	2 (1.81%)
No. of households with farming and four more occupation	7 (26.92%)	4 (9.30%)	4 (23.53%)	0 (0.00%)	15 (13.63%)
No. of households without farming and one more occupation	8 (30.77%)	7 (16.28%)	0 (0.00%)	0 (0.00%)	15 (13.63%)
No. of households without farming and two more occupation	6 (23.08%)	1 (2.33%)	1 (5.88%)	1 (4.17%)	9 (8.18%)
No. of households without farming and three more occupation	1 (3.85%)	2 (4.65%)	0 (0%)	0 (0%)	3 (2.72%)
Total	26	43	17	24	110

Source: field survey 2012.

combined figures appear to be 30.77%, 60.47%, 64.7% and 70.83% for Tant Khanda, Baje Salepur, Jotram and Kandorsona, respectively. If pure non-farm is merged with this combined dependence then the percentages stand as 88.47%, 83.73%, 70.59% and 75% respectively. Thus, there is a clear view that the proportion of households pursuing only farming is quite small due to low returns from farming.

Households falling in category (ii) are further classified into four more groups depending on the increasing number of non-farm activities they combine with farm activity. Except in the case of Tant Khanda, we see a great number of people pursuing one or two non-farm activities together with farm activities, and a few pursuing even more. In Tant Khanda we find an exception where some people even combine four more non-farm activities with farming.

Similarly, we classify those households solely dependent on the non-farming sector on the basis of the number of non-farm activities they are undertaking. In Tant Khanda and Baje Salepur, we found a great number of people dependent on a single non-farm activity. In Jotram and Kandorsona we find only one case for each under this category and they are diversifying into two non-farm activities.

6. Determinants and impact of diversification

The analysis so far shows that most of the households tend to diversify their income sources. However, the results are descriptive in nature. Thus, to identify the major factors influencing the Livelihood Diversification Index (LDI) a multiple regression analysis is carried out using equation (1), where DI is the explanatory variable representing the Livelihood Diversification Index and AGE, EDU, FSIZE, LOAN, ASSET, LVSTCK and DR are the explanatory variables which are defined below.

$$DI_i=a_0+a_1(AGE)_i+a_2(EDU)_i+a_3(FSIZE)_i+a_4(LOAN)_i+a_5(ASSET)_i+a_6(LVSTCK)_i+a_7(DR)_i+e_i \ldots\ldots\ldots(1)$$

6.1 Descriptions of the explanatory variables used in the regression analysis

Variable name	Definition
AGE	Represents the age of the family member engaged in non-farm activity
EDU	Education is represented by the years of schooling of the person engaged in a non-farm activity.

FSIZE Family size means total number of members in a household.
LOAN Stands for the amount of credit/loan taken by a household during the last five years
ASSET This stands for estimated value of all the physical assets(including land) owned by a household
LVSTCK This represents the estimated value of livestock possessedby a family.
DR Dependency Ratio, defined as the ratio of the number of non-earning members to the number of earning members of a family.

6.2 Rationale behind choosing the above variables as explanatory factors

Age

Age is expected to be an important factor affecting the level of diversification. The younger a person, the greater their likely zeal to try different avenues of earning. With increasing age, people like to adopt specific job areas with which they develop a liking with repeated applications. Therefore, the extent of diversification is expected to be negatively related to the age of the person engaged in a non-farm activity.

Education

It is expected that as a person will be more educated they will become more eligible for the high-end non-farm sector, and will also be more aware of potential employment and earning from different non-traditional areas. Therefore, level of education should have a positive impact on diversification.

Family Size

The general idea is that with the increase in family size, if the number of children and elderly people increases in a family, the dependency ratio increases, and this will motivate farmers to search for alternative sources of income. We can therefore expect the family size to be positively related to diversification.

Loans

The impact of loans on diversification will depend not only on the amount of a loan but also on the purpose of taking it. If it is taken for starting up a new enterprise or for purchasing a productive asset, it should affect diversification positively. However, if the loan is used for unproductive purposes like household expenditure, medical expenses, daughter's marriage etc., then it will have little effect on diversification.

Asset value

In asset value we have included both agricultural and non-agricultural assets. From the literature we found that agricultural prosperity strongly influences diversification towards non-farm activities (Jeemol Unni 1991). Now, increase in agricultural asset value can be taken as an indicator of agricultural prosperity. Therefore, we can expect that increase in agricultural asset value will affect diversification directly. On the other hand, non-agricultural assets include public media (radio, television) along with others. Listening to the radio and access to televisions are effective tools in influencing diversification and welfare conditional on other relevant variables. Hence, assets as a whole might affect diversification positively.

Livestock value

Livestock can be linked with either agriculture or it can be taken as an alternative source of income. Therefore, presence of livestock represents either their intensive utilization in agriculture or as an alternative source of income, and might also reduce the tendency to diversify. Thus we can relate livestock value negatively with diversification. Again, from the literature we have found that there is a strong linkage between agricultural prosperity and diversification (Jeemol Unni 1991),and as increase in livestock value can be taken as a reflector of agricultural prosperity, it is therefore implied that there might exist a positive relation between livestock value and diversification. So, the relation between livestock and diversification is likely to be ambiguous.

Dependency ratio

As the number of dependent persons increases, the pressure on the earning members of a family increases in order to fulfil its basic needs, and therefore they try to find more and more avenues of earning and tend to diversify. Hence, the extent of diversification can be hypothesized to be positively related to the dependency ratio.

Table 17.4. Determinants of household non-farm diversification - regression result

Explanatory Variables	Villages				
	Tant Khanda	Baje Salepure	Jotram	Kandorsona	Combined
Age	-9.89E-03**** (-1.152)	1.314E-02 (1.202)	-1.09E-02 (-0.654)	-1.71E-02*** (-1.685)	-5.09E-03 (-.968)
Education	-3.23E-02** (-2.097)	2.454E-02 (.794)	2.532E-02 (.447)	-4.09E-03 (-0.124)	3.23E-03 (.242)
Family Size	8.666E-03 (0.099)	0.1*** (1.683)	0.241*** (1.810)	2.259E-02 (.225)	8.582E-02* (2.410)
Loan	6.771E-08 (.366)	-2.66E-06*** (-1.826)	1.398E-07 (0.164)	-4.14E-07 (-0.307)	-2.59E-07 (-.904)
Asset Value	-5.36E-07**** (-1.305)	1.019E-07 (0.189)	-1.37E-07 (-0.556)	-1.66E-07**** (-1.438)	-1.15 E-07**** (-1.601)
Livestock Value	-4.08E-06 (-0.717)	1.443E-05**** (1.530)	1.93E-07**** (-1.422)	-362E-05*** (-1.945)	1.257E-07 (.248)
Dependency Ratio	0.590* (5.283)	0.382* (3.758)	0.33** (2.622)	0.526* (3.694)	0.417* (8.139)
R^2	0.839	0.424	0.869	0.571	0.513

Note: Figures in parenthesis indicate the t values. * and ** indicate level of significance at 1% and 5% respectively

6.3 Interpretation of the regression results

From Table 17.4 above we see that the R^2 values of our model are quite satisfactory in all of our cases. Therefore, we can say that our model is a good fit. Let us analyse the regression results by considering the independent variables one by one.

Age: The co-efficient is positive but insignificant for Baje Salepur, whereas it is negative for the rest of the cases, though insignificant for Jotram and all villages taken together. Therefore, according to the expectation that age should be negatively related to diversification, our results in four of five cases have supported our expectation.

Education: The co-efficient is positive for Baje Salepur, Jotram and all villages taken together, though insignificant in all cases. On the other hand, it is negative for Tant Khanda and Kandorsona, though insignificant for Kandorsona. According to our expectation, education should affect diversification positively. Though theoretically diversification should increase as education increases, we have experienced that the kind of jobs which people of our surveyed area have opted for basically fall in the category of non-agricultural labour and petty trading, for which education is not an important determining criterion. The higher-educated people we have interacted with are mostly service holders and high salaried persons who are not much interested in diversification. Therefore, the above negative result is very much expected in the context of our study, which contradicts the traditional idea.

Family size: The co-efficient is positive in all of our five cases, though insignificant for Tant Khanda and Kandorsona. Therefore, according to our line of thinking, family size is positively related to diversification. The possible explanation may be that increase in the number of family members induces people to diversify their sources of income.

Loan: The co-efficient is positive but insignificant for Tant Khanda and Jotram, while for Baje Salepur, Kandorsona and all households taken together the co-efficient is negative, though insignificant for Kandorsona and the combined data. This portrays a very weak and ambiguous relationship between loan and diversification. This may be because many of our surveyed households who have taken loans have actually used them for unproductive purposes. As a result it failed to make any significant impact on diversification.

Asset value: Here the co-efficient is negative for Tant Khanda, Jotram, Kandorsona and all households taken together, though for Jotram it is insignificant. On the other hand, the co-efficient is positive but insignificant for Baje Salepur. This shows a result opposite to our prior expectations. This may be because for our surveyed households, most of the asset value comprises agricultural assets. Almost 90% of total asset value comprises the agricultural assets (see Table 17.4 above) which indicates concentration towards agriculture. Therefore, a negative relation between asset value and diversification is obvious.

Livestock value: For Tant Khanda, Jotram and Kandorsona the co-efficient of livestock is negative, though for Tant Khanda it is insignificant. On the other hand, for Baje Salepur and all the sample households taken together the co-efficient is positive, though for the combined data set it is insignificant. This reflects an ambiguous connection between livestock value and diversification as we have already expected in our hypothesis.

Dependency ratio: In all five cases we get a highly significant positive relationship between the dependency ratio and diversification. Therefore, our expectation that the dependency ratio affects diversification positively is highly supported by our data.

6.4 Constraints to livelihood diversification

Though non-farm diversification is an important survival strategy for the rural households in the developing countries, there are several hindering factors to successful non-farm diversification. This study has identified some of the socio-economic, technological, institutional and policy constraints to livelihood diversification.

Lack of Credit Facilities: Lack of access to institutional credit for people with a low level of assets is one of the detrimental factors to diversification in the study area. In the absence of credit support from the institutional agencies, the resource poor households are not able to start their own non-farm business or enterprises. Many households in the study area reported that after completion of training provided by the private or government agencies on some self-employment activities, they could not start their own business due to lack of finance/credit.

Lack of Awareness: Rural households in the study area are often unaware of schemes provided by the government for the development of the rural sector, and as such remain outside the purview of getting benefits from the schemes.

No new avenues of employment opportunities: Opportunities for non-farm jobs, within or around the sample villages, are very low. Therefore, the households do not have much scope to diversify their livelihood portfolio.

Lack of Entrepreneurship: Lack of entrepreneurial skill and ability to take risk is one of the main constraints in starting new initiatives or self-employment projects.

To sum up, the principal constraints faced by the rural households in the study area are of different kinds, and most of them are socio-economic in nature and can be overcome if proper government initiatives are taken. Though the Government of India is running many wage employment programmes, however, the performance of these schemes is not encouraging. Many of these programmes have little relevance to local resources, needs and priorities, and the greatest hurdle is the corruption prevailing in the development programmes, particularly in implementing the MNREGA programme. The constraints in the implementation of the programmes are ignorance, illiteracy and indifference of the rural people towards the development programmes, lack of resources, wrong selection of beneficiaries, local dominant culture, factionalism, lack of identification of the needs of local people, etc.

6.5 Impact of diversification on living standard

The well-being of a household can be assessed on the basis of socio-economic parameters like type of dwelling units, educational status, source of drinking water, domestic possessions, mobility status etc. In our analysis we have used total family income (TOTINC), toilet type (TOITYP) and the Diversification Index (DI) as the independent variable to explain the perceived standard of living of the surveyed villagers. A logit regression has been applied for the analysis where the dependent variable standard of living (LVNSTNDRD) is qualitative and binary in nature, taking value 1 when improvement in living standard is perceived and value 0 when no such improvement is perceived.

6.6 Logit Regression Analysis

To identify the major factors influencing the standard of living of the people in our study area, a binary logit regression analysis is carried out using equation (2).

$$(LVNSTNDRD)_i = a_0 + a_1\,(TOTINC)_i + a_2\,(TOITYP)_i + a_3\,(DI)_i + e_i. \qquad \ldots\ldots\ldots\ldots (2)$$

6.7 Description of the explanatory variables used in the regression analysis:

Variable name	Definition
TOTINC	Monthly income of a household from different sources.
TOITYP	Type of toilet of a household (kachcha /pacca). The dummy variable is constructed by taking 1 for a pucca toilet and 0 for kachcha toilet.
DI	Diversification Index calculated by the inverse of HH index.

6.8 Rationale behind choosing the above variables as the explanatory factors

Total income: After fulfilling the basic needs, as income rises, a household concentrates on its standard of living. They try to make their life more comfortable with their extra earnings. Therefore, it is expected that standard of living increases with the increase in income beyond the subsistence level of earning.

Toilet type: Toilet type is a very important indicator of standard of living. It reflects how much a household is conscious about the health and hygiene. In the village area we find several such cases where there exists a pucca house with a kachcha toilet. Therefore, to capture the actual picture of standard of living, the toilet type is no less important than the housing type.

Diversification index: Level of diversification is expected to raise the level of income of a household which is also supported by our linear regression analysis, and level of income influences the standard of living positively. Therefore, it might be hypothesized that diversification will affect the standard of living positively.

Table 17.5. Logit estimates of the impact on household standard of living (perceived)

Explanatory Variables	Villages				
	Tant Khanda	Baje Salepure	Jotram	Kandors ona	Combin ed
Total Income	6.29E-05* (5.375)	1.883 (0.182)	3.553E -08 (0.003)	4.995 (0.707)	1.09E-05** (2.402)
Toilet Type	0.120 (1.070)	-0.001 (-0.004)	0.256 (0.687)	0.253 (0.858)	0.168**** (1.415)
Diversification Index	.425* (5.035)	0.120** (2.098)	-0.058 (0.368)	0.118 (1.144)	0.174* (3.308)
R^2	0.756	0.117	0.067	0.161	0.167

Note: Figures in parenthesis indicate the t values. * and ** indicate level of significance at 1% and 5%, respectively

6.9 Implication of the regression results

From Table 17.5 above, we find that the R^2, except in the case of Tant Khanda, possesses a very low value in the remaining cases. This implies that we have not considered many other factors which are responsible for changes in the standard of living. Incorporation of more relevant variables influencing the standard of living might have increased the R^2 value of the model. Total income is affecting the standard of living in all five cases and the co-efficient is significant in two cases (Tant Khanda and pooled data). Toilet type of a household is affecting the standard of living positively in all the cases except Baje Salepur (though the co-efficient is insignificant) and the co-efficient is significant in one case only (pooled data). Except for the case of Jotram, the diversification index shows a positive influence on standard of living in the remaining cases while it is significant for three cases (Tant Khanda, Baje Salepur and the pooled data). Therefore, we can conclude that all of our variables have certain positive effects on the perceived standard of living of the households.

7. Conclusion and policy suggestions

The study reveals that though the area is agriculturally advanced, 75.45% people of the area are engaged in non-farm activities and 80.44% of total income of the area comes from the non-farm sector. Again, among different categories of non-farm income, business alone shares 47.62% of the total income, proving the high degree of income diversification of the rural households towards non-farm activities. Among the influencing factors apart from diversification, our study indicates the important role of individual's age, family size and the dependency ratio of a household. Younger people are adventuring more into the different avenues of non-traditional non-farm income sources than their older counterparts. Households with higher family sizes and higher dependency ratios tend to diversify more. On the other hand, we have also found some hindering factors to non-farm diversification, such as lack of formal credit facilities, lack of awareness about different employment schemes provided by the government, lack of new avenues of employment, and lack of entrepreneurial skills. However, the study reveals that the impact of diversification on the standard of living of the rural households is positive though not substantial. As long as the hindering factors exist, the benefits of diversification will be limited.

The future non-farm development strategy should thus be aimed at promoting high-return activities and creating employment opportunities for the low income households, making local institutions more effective. Still, there are many constraints that impede the process of diversification. Therefore, policies that reduce constraints to diversification and widen its possibilities are generally desirable. Enabling and facilitating the environment for the spread of diverse non-farm income-generating activities can be a solution especially for poor households.

- The average for years of schooling of the people in the area is around 7 to 8. With this level of education, our data show no impact of education on diversification. It may be the case that this level of education is not useful in penetrating the high-value non-farm job market. On the other hand, in the low-value non-farm job market, where experience rather than education is the more important criterion, they are lagging behind due to proper training. Therefore, by encouraging people towards higher education and by making complementary investments on community level basic skills training programmes, substantial progress can be made in the direction towards diversification.
- Distant market and related transport costs are the major factors deterring people from using markets. Focus should be put on improving marketing access to them. Special attention in this regard should be given to establishing more local cooperatives and creating linkages with urban areas as well as the international market.
- The fact that family size inhibits the chance of highly diversifying indicates that population pressure is of important concern in the area. There is a substantial Muslim population in the area, and average family size is relatively higher for them than their Hindu counterparts. Therefore, awareness creation and provision of family planning services are mandatory in the region.
- Petty trade was found to be the major item in the non-farm income share of households. Provision of technical support and developing linkages to mainstream financial institutions is necessary to initiate an entrepreneurial culture and business.
- Since the area has a good potential for irrigated agriculture development, small-scale, locally owned irrigation schemes need to be emphasized for enhanced non-farm diversification.

- In the area a 100-days employment scheme is running under the provision of NREGA, but a proper utilization of the scheme is not being made. The Panchayats should ensure at least 100 days work.
- Rural credit facilities on easy terms are available to the villagers, which might impact their standard of living and employment opportunities.
- The tendency and pattern of non-farm diversification is far from homogeneous. For instance, youngsters are more diversifying than the older people, and women are participating in petty trades. Therefore, policy makers need to devise different strategies for different groups of society, and they must be governed with a "one size does not fit all" philosophy.

References

Barrett, C. B., Reardon, T. & Webb, P. "Non-Farm Income Diversification and Household Livelihood Strategies in Rural Africa: Concepts, Dynamics, and Policy Implications." *Food Policy* 26 (4) (2001): 315–31.

Bhaumik, S. K. "Diversification of Employment and Earnings by Rural Households in West Bengal." *Indian Journal of Agricultural Economics* 62 (4) (2007): 585–606.

Bryceson, D. F. "African Rural Labour, Income Diversification & Livelihood Approaches: A Long-Term Development Perspective." *Review of African Political Economy* 26 (80) (1999): 171–89.

Chambers, R. & Conway, G. R. "Sustainable Rural Livelihoods: Practical Concepts for the 21st Century." IDS Discussion Paper No. 296, Brighton, U.K., 1992.

Ellis, F. "Household Strategies and Rural Livelihood Diversification." *Journal of Development Studies* 35 (1) (1998): 1–38.

—. *Rural Livelihoods and Diversity in Developing Countries*. Oxford: Oxford University Press, 2000

Haggblade, S., Hazell, P. & Reardon, T. "Introduction." In *Transforming the Rural Nonfarm Economy*, eds. Haggblade, S. P. Hazell & T. Reardon. Baltimore: Johns Hopkins University Press, 2007.

CONTRIBUTORS

Dr. Arpita Banerjee, Assistant Professor, Department of Economics, M.U.C Women's College, Burdwan, West Bengal, India

Debasish Batabyal, Research Scholar, Department of Tourism Management, The University of Burdwan, Burdwan, West Bengal, India

Dr. Swapan Kumar Biswas, Professor, Department of Commerce (Ret.), The University of Burdwan, Burdwan, West Bengal, India

Dr. Namita Chakma, Assistant Professor, Department of Geography, The University of Burdwan, Burdwan, West Bengal, India

Tonmoy Chatterjee, PhD Student, Department of Economics, Rabindra Bharati University, Kolkata, West Bengal, India

Dr. Soumyendra Kishore Datta, Professor, Department of Economics, The University of Burdwan, Burdwan, West Bengal, India

Tanushree De, PhD Student, Department of Economics, The University of Burdwan, Burdwan, West Bengal, India

Dr. Utpal Kumar De, Department of Economics, North-Eastern Hill University, Shillong, India

Dr. Santosh Kumar Dutta, Associate Professor, Department of Economics, Hooghly Mohsin College, Hooghly, West Bengal, India

Dr. Swapna Ghorai, Associate Professor and Head of the Department of Geography, Raja N.L.Khan Women's College, Midnapore, West Bengal, India

Dr. Ambarnath Ghosh, Professor, Department of Economics, Jadavpur University, Kolkata, West Bengal, India

Banani Ghosh, Guest Lecturer, Department of Geography, Raja N.L.Khan Women's College, Midnapore, West Bengal, India

Biswajit Ghosh, Assistant Teacher, Khorad Amena High School, Satghia, Barddhaman, West Bengal, India

Dr. Kausik Gupta, Professor of Economics, Vice Chancellor, West Bengal State University

Notan Bhusan Kar, Independent Researcher, writer, India

Dr. Pravat Kumar Kuri, Associate Professor, Department of Economics, The University of Burdwan, Burdwan, West Bengal, India

Dr. Ruma Kundu, Assistant Professor, Department of Economic Studies and Planning, Sikkim University, Sikkim

Dr. Arindam Laha, AssistantProfessor, Department of Commerce, The University of Burdwan, Burdwan, West Bengal, India

Arup Majumder, Ph.D Student, Department of Anthropology, Vidyasagar University, Paschim Medinipur, West Bengal, India

Debjyoty Mukherjee, PhD student, Department of Economics, The University of Burdwan, Burdwan, West Bengal, India

Anish Kumar Mukhopadhyay, Assistant Professor in Economics, J.N.M.S. Mahavidyalaya, Nahata, West Bengal, India

Dr. Pradip Kumar Parida, Assistant Professor, IIPA, IP Estate, New Delhi, India

Soumya Sahin, Senior Research Fellow, Jadavpur University, Kolkata, West Bengal, India

Dr. Atanu Sengupta, Associate Professor, Department of Economics, The University of Burdwan, Burdwan, West Bengal, India

Nirmalendu Sarkar, Dinabandhu Mahavidyalaya, Reader in Commerce, Bongaon, 24 Pgs(N), West Bengal, India

Soumitra Sarkar, Assistant Professor of Commerce, Alipurduar College, India.

Dr. Suchismita Mondal Sarkar, Assistant professor, Department of Economics, The University of Burdwan, Burdwan, West Bengal, India.

INDEX